SPECIAL FORCES

SNIPER

SKILLS

OSPREY
PUBLISHING

SPECIAL FORCES
SNIPER
SKILLS

Robert Stirling

First published in Great Britain in 2012 by Osprey Publishing,
Midland House, West Way, Botley, Oxford, OX2 0PH, UK
44-02 23rd Street, Suite 219, Long Island City, NY 11101, USA
E-mail: info@ospreypublishing.com

OSPREY PUBLISHING IS PART OF THE OSPREY GROUP

A CIP catalogue record for this book is available from the British Library

ISBN: 978 1 78096 003 6
E-pub ISBN: 978 1 78200 765 4
PDF ISBN: 978 1 78200 764 7

Index by Sharon Redmayne
Typeset in Palatino
Originated by PDQ Media, Bungay, UK
Printed in China through Worldprint
Diagrams by Peter Bull Art Studio

12 13 14 15 16 17   10 9 8 7 6 5 4 3 2 1

Osprey Publishing is supporting the Woodland Trust, the UK's leading woodland
conservation charity, by funding the dedication of trees.

www.ospreypublishing.com

Front cover images courtesy of Getty Images.

Acknowledgements
The author would like to express his appreciation for the advice and assistance given in
the preparation of this book by superlative sniper, and trainer of snipers, Colour Sergeant
Michael Bruce McIntyre BEM of the British Parachute Regiment.

CONTENTS

LIST OF IMAGES

29. A US Marine sniper in mountain warfare training.
30. British Army L115A3 bolt-action sniper rifle.
31. British Arctic Warfare bolt-action rifle.
32. Special Forces weathermen supply vital battlefield information.
33. US Army Sniper School – snipers in training.
34. Sniper team during mountain training.
35. Advanced Combat Optical Gunsight on the range.
36. Sniper's view.
37. Sniping position 1 – lying prone.
38. Sniping position 2 – lying raised.
39. Sniping position 3 – sitting.
40. American sniper team while on operation.
41. A rebel sniper in Tripoli, Libya.
42. US Secret Service sniping unit.
43. A British sniper from the Irish Guards during an operation in Basra.
44. A British sniping team during a strike operation in Afghanistan.
45. A US sniper team in action in Eastern Afghanistan.
46. A US Marine sniper during mountain training.
47. A camouflaged US Army Special Forces sniper.
48. Reoccurring tactics: a US soldier attempts to draw fire using his helmet.
49. A US soldier instructs an Iraqi soldier on the use of the Barrett M107 sniper rifle.

INTRODUCTION

The purpose of this book is to provide quite a detailed introduction to the skills, equipment, tactics and operations that relate to the practice of sniping amongst 21st-century special forces. We start with a brief journey through the origins of sniping, to make the purpose and motivation of the art plain, before moving on to discuss the development of weapons and equipment, particularly over the last two centuries. We finally arrive at an examination of the tactics and training that together produce the modern military sniper, a true elite amongst the world's armed forces.

Today, the armies of most nations feature a special forces element, be that the US Navy SEALs (short for 'Sea, Air, Land'), the Russian Spetsnaz or the British Special Air Service (SAS). Within these formations there are chosen men selected to serve as snipers, those deemed to be equipped both mentally and physically to carry out the sniper's arduous or lonely mission, often a long way from support forces. Though these snipers have several roles, the one thing which ties them all together is precision violence delivered with a long-range rifle. Most of the time, either the sniper will be working from a fixed position and controlling an area around a military base, or he will be moving silently and stalking a human target of some kind, civilian or military. Beyond this, and due to his skills in field craft and movement, a sniper is often used as a forward observer (FO) for artillery control, or even to conduct reconnaissance or gather intelligence. Sometimes, the sniper may be sent out in a counter-sniper role – fighting someone of his own profession – or may be employed in a hostage-rescue situation. He might even be assigned to kill the enemies of his country in a covert operation overseas under the command of intelligence operators.

Regardless of the type of operation, the devil is in the detail. The basis of all sniper training, though not all operations, is that the sniper should be able to remain hidden from the enemy both while moving across country and while concealed in his shooting position. Without the skills of camouflage and concealment, a sniper loses the initiative and renders himself as vulnerable as any other infantryman. It is only when the sniper is thoroughly capable of exercising high level skills in field craft and tactics that he may get the opportunity to use his shooting skills to destroy an enemy at more than a mile.

The most important difference between a sniper and any other soldier relates to his mental makeup. Of course, he requires a high level of courage, field craft and shooting skills, but so do most fighting soldiers. What is different about a sniper is that he is emotionally capable of killing in cold blood, and after close observation of his target. He may, for instance, have been watching a target shaving and cooking for a day or two before killing him. This is a very different proposition to shooting a man in the frenzy of open combat. Neither is there, most of the time, any hurry to make a split-second decision. The sniper has plenty of time to think about what he is doing and either kill or allow the target to escape.

Not everyone can operate like this and remain a useful soldier. One interesting thing about sniper selection is that most people know in advance if they can do this or not. All snipers learn pretty much the same skills but, of course, some snipers are taught better and some learn better depending on the individual and their army's training programme. Much of the time special forces snipers work on similar missions to infantry of the line, but perhaps in particularly violent or environmentally difficult areas. Some of the missions they are given might surprise us – 'Black Ops' can be very black indeed. Sometimes, in the fog of both politics and war, it is hard to tell the good guys from the bad.

The attitude to snipers on the battlefield has changed over the last 10 or 20 years, especially as counter-insurgency warfare in Iraq and

Afghanistan has become a major news item. Before 2001, snipers were often hated by soldiers everywhere as cold-blooded murderers – even those on their own side might think they were butchers. Many young people might find that attitude surprising, given the heroic image of snipers today. Any reader who is under about 40 years old will have grown up in a different era, with different ideas and values to older veterans. In the 'old days' of World War I and World War II, armies were made up of conscripts who were in combat because they had to be; men who were trying just to stay alive and keep their comrades alive in the process. A professional sniper who was deliberately going out of his way to sneak up on and kill an enemy – an enemy who was often just other ordinary soldier – was therefore considered by most soldiers to be a little short of a murderer, even if the sniper was on their side.

Of course, snipers on the battlefield then did a very different job to what snipers are doing today. The job of the sniper in the past was to stalk and kill high-value targets such as officers, senior NCOs, radio operators and so on if they could find them, but the guy making the tea, an ordinary conscript soldier, if they couldn't. Today things are very different. The US Army's 1994 *Sniping Training* field manual touches on the broad demands of sniping in its introduction:

The primary mission of a sniper in combat is to support combat operations by delivering precise long-range fire on selected targets. By this, the sniper creates casualties among enemy troops, slows enemy movement, frightens enemy soldiers, lowers morale, and adds confusion to their operations. The secondary mission of the sniper is collecting and reporting battlefield information.

a. A well-trained sniper, combined with the inherent accuracy of his rifle and ammunition, is a versatile supporting arm available to an infantry commander. The importance of the sniper cannot be measured simply by the number of casualties he inflicts upon the enemy. Realization of the sniper's presence instills fear in enemy troop elements and influences their decisions and actions. A sniper enhances a unit's firepower and augments the varied means for destruction and harassment of the enemy. Whether a sniper is organic or attached, he will provide that unit with extra supporting fire. The sniper's role is unique in that it is the sole means by which a unit can engage point targets at distances beyond the effective range of the M16 rifle. This role becomes more significant when the target is entrenched or positioned among civilians, or during riot control missions. The fires of automatic weapons in such operations can result in the wounding or killing of noncombatants.

b. Snipers are employed in all levels of conflict. This includes conventional offensive and defensive combat in which precision fire is delivered at long ranges. It also includes combat patrols, ambushes, countersniper operations, forward observation elements, military operations in urbanized terrain, and retrograde operations in which snipers are part of forces left in contact or as stay-behind forces.

– US Army FM 23-10, *Sniping Training*, 1-1 (1994)

While paragraph (a) reflects the classic sniper model, paragraph (b) hints at the sniper as a true multi-tasking individual, switching between different types of combat skill across a varying landscape of warfare. In a modern counter-insurgency situation the task of a sniper, on the side occupying the country and suppressing the insurgency, is not to seek out high-value targets in a target-rich environment so much as to pick out single insurgents from amongst a population of civilians. In conventional warfare, the soldier knows where the frontline is and who the enemy are. Yet in counter-insurgency warfare, the enemy are all around and hidden amongst the civilian population, whom the occupiers cannot shoot for obvious reasons. Although the sniper can kill with clinical efficiency, his job has become far more complicated. The responsibility resting on his shoulders is frequently a heavy burden, one requiring the ultimate levels of training and mental composure.

CHAPTER 1
THE EVOLUTION OF PRECISION FIRE

The modern sniper is the culmination of a long evolution in precision-fire technology, from the bows of ancient history through to modern sniper rifles with their precision optical sights. Understanding some key points of this evolution helps us to comprehend not only the tools of the sniper, but also something of the tactical opportunities and constraints inherent in his role.

Looking back to ancient history, we find direct forerunners of the sniper, particularly in the various types of archer utilized by the world's armies. Well-trained archers, like snipers, were an elite of their day, and were individuals of extraordinary talent. Assyrian archers of the 2nd millennium BC, for example, could hit a single human target with an arrow at several hundred metres, and from horseback, using their composite bows (bows made from a mixture of wood, horn, bone and sinew, with improved elasticity over pure wooden 'self' bows). The concept of precision fire was therefore practically, if not theoretically, established by the time of the Greeks and the Romans.

The apogee of early precision infantry fire was arguably the medieval longbow. The longbow was the most powerful and effective weapon on the Western European battlefield for the 200 years between 1250 and 1450. In the right hands it decided battles and changed history. The Welsh longbow, as I shall refer to it here, is so called because the Welsh are the first people recorded as using it. The same bow is also known as the English longbow because the

English deployed it on the battlefield repeatedly with great effect. Yet it was used with far less effect by armies on continental Europe. (Why this was so, we shall see in a moment.)

The longbow was the ultimate projectile weapon before the use of firearms. (In point of fact, in the hands of a trained archer it is far more effective than a musket.) It was capable of launching more than nine arrows a minute for brief periods and six arrows a minute for an extended period, and in trained hands it could hit a man at 150m (164yd). Compare that with the maximum of four rounds per minute from a musket accurate individually to about 50m (55yd) and in volley fire out to only 100m (109yd). Volley fire from the longbow was capable of killing men and horses out to 450m (492yd). At any range a suitably tipped arrow – a bodkin-spike for mail armour and a type of blade-tip for horses and other armour – would kill a man or horse. The only thing that would stop such an arrow was the heaviest, highest quality plate armour available to only the richest knights after about 1350. Even then, his horse would be killed under the rain of arrows, as a horse could never carry sufficient armour to withstand the armour-piercing arrow head. In short, a line of English or Welsh archers could stop any armoured cavalry charge in its tracks or kill enemy infantry or crossbow-men before they were close enough to do damage.

During the Hundred Years War, between the English and the French, the Welsh longbow was a major contributor to victory at the battles of Crecy in 1346, Poitiers in 1356 and, most famously, the battle of Agincourt on Saint Crispin's day, 25 October 1415, which crippled the French Army and allowed English King Henry V to marry the King of France's daughter, Princess Catharine of Valois.

The Welsh longbow was, and still is, made from a piece of yew tree. For many years it was thought that the medieval longbow was 1.5-1.8m (5–6ft) in length, but the large number of preserved longbows found on the wreck of the 16th-century ship, the *Mary Rose*, were all between 1.85m (6ft 1in) and 2.11m (6ft 11in) long. In order to prevent warping later, the yew wood was gently dried for about two years before it was cut and carved to its final bow shape. To form the longbow, the suitable length of wood was shaped into a 'D' section in such a way that the inner side of the bow was formed of heartwood and the outer flat side was formed of the softer sapwood. This arrangement, which demanded a great deal of skill to manufacture, took advantage of the differing elasticity of the two types of wood under compression and extension. When ready for action, the bow was strung with a string of hemp, flax or even silk, if the archer could afford it. When not in use, however, the bow was unstrung to preserve its strength.

The longer the bow, the more energy it could release over a longer length of arrow, and the thicker the wood the greater the strength required to draw the bow and the greater the energy imparted to the arrow when fired. A modest bow, which would take a normal but very strong modern person all their strength to draw, might require a pull of 34kg (75lb) to full stretch, yet some medieval bows were made with a pull requirement of more than 68kg (150lb). And this brings us to why the English deployed the Welsh longbow with such great effect in the European wars during the Middle Ages, and no one else did – the English had plenty of skilled bowmen.

As we have seen, the longbow was very difficult to draw and even more difficult still to aim under tension. Even a 'learner' bow takes a strong man all his effort to draw, until the right muscles are built up through regular exercise and practice. It therefore took many years of steady effort and unremitting practice to make a good bowman. With this is mind, the English kings brought in a law that required all boys and men of fighting age to practise every Sunday with the longbow. To make this practice popular, competitions were arranged and prizes were awarded, and a good bowman was paid a very good wage when on a military campaign. This legal and financial encouragement of bowmanship paid huge dividends in England's many medieval wars.

As well as delivering volley fire to a broad area, there are instances in which the longbow was used for more precision 'sniping'. The first confirmed kill of a selected high-value target in a military conflict was when the Welsh, armed with their trusty longbows, were fighting the Northumbrians in 1633. On 12 October the Northumbrians took the field in marshy ground on the banks of the River Don about 19km (8 miles) north-east of the modern town of Doncaster in the county of South Yorkshire. They were commanded, at what became known as the battle of Hatfield Chase, by their king, Edwin, who was known as a seasoned warrior and good tactician. The king was supported by his son Prince Offrid and a number of seasoned nobles. In support of his main infantry force, King Edwin had his archers strung out towards Doncaster to harass the approaching Welsh and Mercian alliance, formed by the King of Mercia, Panda, who appears to have been jealous of Edwin's success in uniting his kingdom.

Despite the best efforts of the Northumbrian bowmen, the Welsh approached within bowshot of the English by about lunchtime and charged to cover the ground more quickly and avoid arrow wounds. As they approached a single arrow from a Welsh longbow seems to have been fired at King Edwin, but it missed and killed his son Offrid standing beside him. This loss caused King Edwin to lose his senses and charge alone into the press of Welsh warriors, not caring if his men were with him or not. Very quickly he was speared many times and died on the spot. So, indirectly, this appears to be the first recorded English battle decided by precision fire. At the siege of Abergavenny, Wales, in 1182, Welsh archers hit William de Braose, 4th Lord of Bramber, the English commander. The arrow went through his chainmail, through his thigh, through the saddle and penetrated the horse he was riding. William de Braose was out of action for a while as a result, but not killed.

The Age of Firearms

While bows, as we have seen, were capable of feats of precise long- range killing, it was the advent of firearms that truly laid the path to the modern sniper. The 14th century saw the first appearance of primitive firearms on the field of battle in Europe. By about 1450 a weapon called the harquebus – something like a small-bore, muzzle-loading cannon supported on a vertical stick – started to appear. It was issued to selected infantry and placed amongst other infantry armed with pikes. The tactical idea was that 1,000 men could be trained to use these weapons in a few weeks rather than the lifelong practice required of the longbow archer. Besides the effect of the heavy lead shot they fired, the bang doubtless terrified the enemy horses. But the rate of fire was slow, hence the men with pikes were still needed to defend the 'gunners' from cavalry or a swift infantry charge.

As the price of gunpowder reduced, and the technical men of the day learned to make it to a better standard, the musket, effectively a lighter harquebus, made its appearance and gradually came to dominate the battlefield as the main infantry weapon. A musket is a muzzle-loading, shoulder-supported long firearm originally fired by means of a glowing taper applied to a gunpowder-primed touchhole (given mechanical efficiency in the matchlock mechanism), and later by the spark from a flint striking on iron, most effectively via the flintlock mechanism. All European infantry were armed with muskets by about 1700 and most of these were flintlocks. In the final development of the musket in the early 1800s, the main charge was fired by the detonation of a percussion cap struck by a hammer in the place of the firing pan and touch-hole. This weapon was quicker to load and, unlike the flintlock, it was untroubled by the effects of wind and rain on the priming powder.

All muskets were originally smoothbore, which is one of their defining features, and fired a heavy lead ball that was made significantly smaller than the half or three-quarter inch calibre of the weapon. This was so that the rapid build-up of ash or carbon from burning the dirty gunpowder of the time would not prevent the gun being reloaded. Of course, this loose fit of the ball made the weapons very inaccurate and a proficient user would only be able to hit a man at 50–70m (55–76yd). Not only did the musket have a short range, but it also had a slow rate of fire – something in the region of 4 rounds per minute (rpm) was considered good going for trained infantry. To make best use of the musket's hitting power, infantry would normally line up and fire at groups of the enemy with volley fire, as this would both ensure a reasonable hit rate and keep the soldiers calm and steady while reloading 'by the numbers'. When the opposing infantry came close, firing ceased and the bayonet came into play. The first bayonets were plug bayonets, which were attached by ramming a bayonet-attaching rod down the end of the muzzle and prevented any further firing. Thus the musket effectively became a club with a spike on the end. Only later was a bayonet developed which clipped to the end of the muzzle, parallel but offset, and allowed firing to continue.

Although the standard smoothbore musket was a rather inaccurate weapon, this is not to say that all early firearms were inefficient at range. By the late 1400s someone had the idea of cutting spiralling grooves, called rifling, down the length of the inside of a musket barrel to cause the bullet to spin as it was fired. This initial idea was improved in 1520 by August Kotter, an armourer in Nürnberg, Germany.

The idea of rifling may have come from the archery trick of twisting the flights to make the arrow spin and fly true, by virtue of centrifugal stability. The immediate effect of rifling a barrel was greater range and accuracy, but the primitive technology of the time prevented widespread production of muzzle-loading, rifled weapons even if they had been wanted by the military. By the early 1700s, as production technology advanced, the application of rifling was becoming widespread with hunting enthusiasts and the use of an elongated bullet was giving far greater range and accuracy than the usual spherical ball. Yet the initial refusal of the military to adopt the rifled musket was not as stupid as one might think, owing to several factors. The gunpowder of the day was very dirty in use and quickly fouled the rifling, leading to loading difficulties; no soldier was able to clean his barrel in the middle of a battle. This meant the first few shots would be accurate; thereafter the infantry might not be able to fire at all. Furthermore, the necessarily tight fit between bullet and rifled bore meant that rifles had a slow reloading time compared to smoothbore firearms, decreasing the number of volleys that could be delivered.

Despite the problems, some armies saw the benefit of the rifled musket, and by the late 1700s had formed specialized infantry units comprising, in the British Army for instance, 'chosen men' to test the new weapon in battle. The men were effectively the first special forces soldiers, and were picked for their ability to think and act independently and to grasp the new technology. Across the Atlantic, an early American militia, active during the 1750s, was 'Rogers' Rangers', from whom several Canadian and American units including the US Army Rangers claim descent. These militiamen were an independent company of Americans who fought for the British in North America during the Seven Years' War (1756–63) against the French and Indians. They were organized by Major Robert Rogers into a form of irregular light infantry, and used in scouting, ambush patrols and operations against distant targets. In 1757 the Rangers even fought the first battle on snowshoes, in which they defeated a larger force of French near Fort Carillon at the south end of Lake Champlain.

The Rangers also became very highly valued by the British for the collection of intelligence. During the American Revolution Major Rogers is said to have offered his services to George Washington, who refused him on suspicion of spying. Incensed at this, Rogers turned again to the British. Not all his men went with him, however, and his Ranger soldiers fought on both sides during the War of Independence (1775–83).

Amongst the American forces during this conflict were a great many hunters and woodsmen whose skills in field craft and hunting-rifle marksmanship appear to have left a marked impression on the British they were fighting, and may have led, indirectly, to the formation of British rifle units, particularly the Experimental Corps of Riflemen created by the British Army in 1800. The recruits for this new corps were hand-picked from other infantry regiments, dressed in 'rifle green' as an early attempt at standard-issue camouflage, and armed with the Baker rifle (see below) and a 61cm (24in) sword bayonet. The idea was that, in a pitched battle, the riflemen would skirmish ahead of the main body of infantry and force the enemy to form squares for protection, which made them good field artillery targets, or at least slow their advance as they stopped to return fire. It was also thought they could be used for ambushes and other small-unit operations. Effectively, these new rifle units were the SAS of their time. Within four months of its first parade, the Experimental Corps of Riflemen led the disastrous British landing assault at Ferrol on the north-west coast of Spain. This action nevertheless enabled them to show their abilities in a favourable light and two months later they lost the title 'Experimental' and were gazetted under the name The Rifle Corps. Colonel Coote Manningham, who had commanded at Ferrol, was given command of this new unit and the famous Light Infantry was born.

The British Army had learned from the Americans how useful a weapon a rifle could be, particularly in skirmishing and irregular warfare, and the added range and accuracy of the Baker rifles were a marked advantage. The British Board of Ordnance held a competition at the Woolwich Arsenal on 22 February 1800 to find the best rifle design available. It was won by a London master gunsmith named Ezekiel Baker and his design, of course, was the Baker rifle. Now the British had a weapon fit for long-range shooting. The 4kg (9lb) flintlock rifle fired a .615in spherical lead ball at a rate of slightly more than 2rpm and was accurate to a range of some 91-274m (100-300yd) – up to six times greater range than a smoothbore musket – depending on the care taken in loading and the marksman firing it. It was effectively a sniper's weapon, being quite able to pick off individual soldiers in an ambush or skirmish.

The new rifle regiments, once equipped with the Baker rifle and having gained some combat experience, were found to be highly effective when deployed in advance of the main British line of infantry. Spread out in skirmishing lines, which presented difficult targets for the volley fire of French muskets, they would approach the standing French line and shoot the NCOs and junior officers from the kneeling or prone position. The tactic was little different, in principle, from the use of snipers in the world wars, aside from their not using cover. The battle at Cacabelos, northern Spain, on 3 January 1809 illustrates their capabilities. The battle was fought between British and French units over possession of a road and key bridge controlling the line of retreat for General Sir John Moore's army. At one stage of the clash, the British had a section of artillery on the bridge and the French cavalry formed up to charge across the bridge and take them. The cavalry were led in person by the dashing young General Auguste-Marie-François Colbert. Rifleman Thomas Plunkett, an Irish soldier in the British 1st Battalion, 95th Rifles, dashed out from the British ranks and onto the bridge. He then lay on his back with his feet towards the enemy, rested his Baker rifle on his toes and shot General Colbert at a range of between 200 and 600m (183 and 548yd), depending upon whose report is given credence. It is said that Plunkett looked back at the unbelieving faces of his comrades and didn't want them to think it was a lucky shot, so he reloaded and shot the trumpet major who had come to the aid of the general. The authenticity of this last exploit is somewhat more questionable, though Plunkett does seem to have shot around 20 enemy soldiers from the top of a convent in Buenos Aires some time earlier, so clearly he was a calm and methodical reloader.

The rifle was also having an effect in naval use at this time. During the Napoleonic conflict, all navies began placing sharpshooters, as early snipers were called, on the platforms high up the masts of their ships – known as 'fighting tops'. One of the main duties of the British Royal Marines at the time was to man these positions and rain down musket and hand-cannon fire onto the enemy decks below when the opposing ships came alongside each other. Of course, the French also placed marksmen in their fighting tops, with tragic effect for the British at the battle of Trafalgar.

In this most famous of engagements, the British Navy – 27 ships under the command of Admiral Lord Nelson – beat the combined Spanish and French fleets, 33 ships under Admiral Pierre-Charles Villeneuve, on 21 October 1805. Nelson led the British line in the *Victory* and first disabled the French flagship *Bucentaure* before moving on to attack the *Redoutable*. Unfortunately, the captain of the *Redoutable*, Captain Jean-Jacques Lucas, had trained his men well in the tactics of throwing grenades and sniping from the masts. The *Victory* and the *Redoutable* became entangled and French sharpshooters poured fire down onto the deck of the *Victory*. In those days, the captain, admiral and all the ship's officers wore splendid uniforms and elaborate hats, which clearly marked them out for the attention of snipers. Nelson himself wore four huge gold stars, denoting various orders of knighthood, on the breast of his uniform so he was not difficult to identify. The battle raged for four hours and both sides fought bravely, but eventually a French sniper shot Nelson from a position some 15m (50ft) above the deck. The bullet entered the top of his left shoulder, cut down through his lung and came to rest at the base of his spine. He remained conscious throughout the battle, but died shortly after the British victory. The bullet was removed from his body and is now on display at Windsor Castle just west of London. Although the battle was a clear British victory – the French and Spanish lost 22 ships while the British lost none – the battle-changing effect of a single talented shooter had clearly been demonstrated.

The same lesson was reinforced in land warfare throughout the remainder of the 19th century. The American Civil War of 1861 to 1865 was the first war in which there were soldiers who actually operated as snipers, going out alone, or in small groups, to seek high-value targets. Their intent was aided by the introduction of the Minié bullet, a sub-calibre, hollow-base projectile that slid easily down a rifled barrel when loaded but which expanded to grip the rifling when fired (see the next chapter for a more detailed examination of bullet development). The Minié bullet had been invented in France in 1849, declined by the French Army and adopted by the British, who bought the patent in 1851 to go with their new Enfield muzzle-loading rifles. Starting at the beginning of the American conflict, issue of the Minié ball ammunition, and suitable rifled muskets with percussion cap ignition, very quickly became almost universal in the armies of both sides. These technological changes brought astonishing improvements in reliability, range and accuracy. Suddenly all infantry, not just a few sharpshooters, could shoot accurately and expect to hit an enemy some hundreds of yards away. And they could reload quickly too – unsupported frontal assaults or infantry charges became virtually suicidal against such ranks. Of thethree million men and boys who fought in this war, roughly 500,000 were seriously wounded and 200,000 were killed, and 90 per cent of these casualties were caused by the Minié bullet fired from a rifled musket.

It was the Minié bullet that allowed some men to work, for the first time, as what the modern reader would recognize as a sniper. Sniper schools were established on both sides and men were trained in the arts of camouflage, stalking and shooting. Some of these individuals became very effective. Union general John Sedgwick was killed at the battle of Spotsylvania Court House in May 1864 by a sniper approximately 730m (800yd) away. He was placing field guns behind his front line and his gunners were taking cover to avoid being shot. Sedgwick waved towards the distant rebel sharpshooters and mocked his cowering gunners, suggesting they ignore the sporadic shots. 'What are you dodging for? They couldn't hit an elephant at this distance!' Seconds after he spoke, a bullet smashed into Sedgwick's face and killed him instantly.

While the talents of the sniper were becoming evident, the tools of accurate long-range shooting were also improving, particularly in Germany in the second half of the 19th century. The Mauser arms company was formed in 1874 in Germany by the Mauser brothers (Wilhelm and Paul) and a gentleman named Samuel Norris from the US arms firm E. Remington & Sons. At this time most weapons were still muzzle-loaders, rifled or not, and the idea of breech-loading was a novelty. The purpose of the new firm was to experiment with the new bolt-action mechanism (established by weapons such as the Dreyse Needle Gun) and the idea of using a firing pin to ignite propellant contained within a unitary cartridge (a cartridge that contained bullet, propellant and percussion cap in one unit). The first Mauser rifle was the Model 1871, which was quickly adopted by most states of the German Empire except Bavaria. Though it might seem a little primitive now, it was a massive advance on a muzzle-loader. Its bolt action offered both a high rate of fire and the ability to reload in a prone position, the latter advantage being critical for future snipers. The cartridge design varied with different models, but was a huge 11 x 60mm black-powder round. (The reason for the large calibre was that black powder has a slower rate of ignition than modern propellants, so the bullet was driven down the barrel more slowly and to develop sufficient hitting power it had to be heavy.)

The last few years of the 18th century saw a bolt-action rifle arms race between the Germans and their perpetual enemies, the French. The German company Mannlicher developed a box magazine to allow a more rapid rate of fire and, most importantly, smokeless powder had been invented. The lack of smoke was by no means the most significant feature of this new propellant. It also burnt far faster than the old black powder and therefore produced far higher velocities, a very good thing from the point of delivering range and accuracy. By 1887 the French had come up with the bolt-action Lebel Model 1886 rifle, which fired a smokeless cartridge and, with a few modifications in place, rendered all older rifles obsolete. The Lebel fired an 8mm calibre bullet at a muzzle velocity of 700m/sec (2,297ft/sec) which is not so much less than the 800–950m/sec (2,624–3,116ft/sec) of a modern rifle bullet. It could reliably hit a man at 400m (437yd) and produce effective volley fire against massed targets up to 1,800m (1,970yd). The one real fault was the slow-to- load tube magazine, which held eight rounds and ran under the barrel.

The German answer to the Lebel was the Gewehr 1888, designed not by Mauser but by a German committee – the German Rifle Commission – from ideas brought forward from a number of sources. Despite the committee approach, the product was a fine bolt-action rifle with all the advantages of the Lebel plus a five-round, clip-feed magazine loading 8 x 57mm rounds. It was the first 'modern' bolt-action rifle with all the main features of a World War II rifle and many of modern bolt-action rifles. Mauser missed the entry time for the German

Rifle Commission effort owing to the death of Wilhelm Mauser, but continued research to produce the Model 1889/90, which proved to be so devastating in the hands of the Boer commandos during the Second Boer War of 1899–1902.

The 1889/90 model had a small bore at 7.65mm, but this allowed the new rimless round to be lighter than other rounds of the time. The rifle's biggest advantage in combat, however, was the 'stripper clip' loading system, in which a metal clip of five rounds was placed above the open bolt of the rifle and pressed down with the thumb. Reloading could be performed in just a few seconds, and therefore generated a far higher rate of fire than any other rifle of the time. Indeed the stripper clip was so effective it was still in use on bolt-action rifles like the British Lee-Enfield well after 1945. Despite its advantages the new Mauser rifle was, strangely, not adopted by the Germans or any other main European power. But it was a success with a Belgian military attaché who did a deal with Mauser to manufacture one model under licence and founded the Belgian arms firm Fabrique National (FN). FN was more successful in its marketing and sold 150,000 rifles to Argentina plus many more to the Ottoman Empire. The huge Argentinean order overwhelmed the FN factory and it had to outsource production to the UK resulting in the founding of British Small Arms (BSA), in Birmingham, England. And, of course, FN then sold a large shipment to the Afrikaner Boers, under the leadership of Petrus Jacobus Joubert.

Joubert was born in South Africa in 1834, a descendant of the French Huguenot settlers. He was orphaned at an early age and made his way to the Transvaal Republic, where he farmed very successfully. He also studied law and became known and respected for his shrewdness. Though his high mindedness held him back somewhat in politics, which valued compromise above principle, he became commandant general of the South African Republic from 1880 to 1900, when he died of peritonitis. Before he died, however, Joubert had the acumen to purchase 25,000 Mausers for £3.00 each and 10 million cartridges for £6.00 per thousand, just before the Second Boer War, a pretty good deal for the Afrikaners by any standards.

His timing was fortuitous. Following the Napoleonic Wars in Europe, the British took over control of the Cape Colony (in which is modern South Africa) and extended their reach north, causing more trouble for the Boer settlers who, in the 1830s and 40s, undertook what they called 'The Great Trek'. They moved a long way further north and established their own republics of Orange Free State, Transvaal and Natal. Yet the British moved north again, claiming land as they went, and ran into the newly re-settled Boers who gave them the First Boer War in 1880. By 1895 the British had bypassed these Boers and established the country of Rhodesia to their north, famous for its emeralds and copper. The next flashpoint came in 1886 when gold was discovered in the Boer lands. The British wanted it and this led indirectly to the Second Boer War in October 1899, when the Mausers came into play.

The Mauser's light ammunition enabled each Boer to carry 60 rounds, in clips of five, on a chest bandolier, and load them much faster than the British-issued single shot rifles. In a determined firefight, the high rate of fire had a devastating effect on the opposition – 100 Boers would be able to fire as fast as 300 British. The rifle itself was lighter than the British equivalent too.

When two units of soldiers are firing at each other, they are generally as much focused on taking cover as trying to shoot the enemy. To this end they are generally lying down, and perhaps showing only their heads to the enemy. Therefore at any significant combat distance there are relatively few hits as a percentage of the rounds fired. The main effect of fire, at least initially, is to suppress the fire of the enemy by forcing them to ground and cover. Such fire superiority is generally achieved by a higher rate of fire, and when it is achieved the dominant side can perhaps advance to decide the issue at close range, which in the 1800s and early 1900s often meant advancing to contact with fixed bayonets.

An interesting fact is that the Boers did not issue a bayonet for use with their rifles. It is said they preferred to stand off and make use of the range to hit the British at ranges of 800m (875yd). In any event, fitting a bayonet to the end of a rifle makes the weapon extremely inaccurate, because it interferes with the shock wave that runs the length of the barrel upon firing. A rifle barrel is designed to be in a certain state of flexion when the bullet leaves the muzzle; the weight of the bayonet alters this dynamic to such an extent that, for instance, an old Lee-Enfield .303in, known for its extreme accuracy, would fire 25cm (10in) low at 100m (109yd) if it had its bayonet fitted. Owing to their shooting and bush craft skills, the Boers became well respected by the British soldier and a custom arose of never lighting three cigarettes from the same match. The idea was that at night the Boer sniper would see you with the first light, take aim on the second light and kill whoever got the third light. This custom became very common in the trenches of World War I and was still around when the author joined the British Army in the 1970s.

The Boers formed Commandos, of roughly battalion strength (750 men), based on all the men of fighting age in each of their towns. The greater part of the Commandos were mounted infantry, though there were artillery units. The hardy Boer farmers made good guerrilla fighters through being brave and highly motivated, skilled horsemen and generally good shots. Their weakness lay in handling large actions, where discipline became a problem as the independently minded Boers didn't like being told what to do by their leaders. Some Commandos would refuse to obey orders or withdraw if they didn't like how things were being run.

The Second Boer War started as a conventional war and, for ease of understanding, can be split into three phases. In phase one the Boers mounted strong raids into Natal, which had already been ceded to the British, and much of the Cape Province, besieging the cities of Mafeking, Ladysmith and Kimberley. As the British sought to relieve these cities, the Boers then won a series of victories at Colenso, Magersfontein and Spion Kop. These successes provoked the British to send in heavy reinforcements, who secured the Cape, Natal and eventually the Transvaal capital, Pretoria, in 1900. At one stage in the conflict British forces were on their way to relieve Ladysmith when the Boers stopped them at the Tugela River with infantry and modern German artillery. The British unsuccessfully tried to outflank them then pushed some men across the river. The obvious next move was to site British artillery on the hill called Spion Kop in the middle of the Boer lines, so that the guns could play over their positions. A British infantry assault was mounted on Spion Kop at night in thick fog and nearly 1,000 British troops fought their way onto the top of the hill. As dawn broke and the fog cleared, however, they found they were only on top of the lowest of three peaks, and they were overlooked by enemy sited on the other two. Immediately Boer snipers, armed with the superb Mauser rifle, began to pick off the British soldiers from some hundreds of metres range. The Boer leadership realized that if the British could capture the other two peaks they would be able to bring up artillery and destroy most of their forces, so they brought their own artillery to bear on the British. Shortly after their arrival, the British were not only being hit by snipers but 10 artillery rounds a minute were falling amongst them. The British infantry, as the reader might imagine, were keen to dig in but there was little earth – only rock and stone – and the British had only 20 shovels and pickaxes between them. The result was a series of trenches no more than 41cm (16in) deep and a totally unrealistic, untenable defensive position. Very soon these trenches were filled with British dead. Eventually the British were forced to concede defeat, due in large part to the shooting of the Boer snipers.

The British high command did eventually learn from their experiences and the actions of the Boers themselves. This led to their commissioning Lord Lovat to form a unit of scouts and snipers called Lovat Scouts. They were commanded by an American adventurer, Major Frederick Russell Burnham, who was made the British Army Chief of Scouts under Lord Roberts.

Burnham fittingly described his scouts as 'half wolf and half jackrabbit'. The scouts were taught not only to shoot well, but to be experts in field craft and military tactics. And they were the first snipers to be equipped with the heavily camouflaged 'Ghillie Suit', a silhouette-diffusing camouflage suit made from strips of cloth, still worn by snipers today. After the Boer War this regiment went on to formally become the British Army's first sniper unit, though they were referred to as 'sharpshooters'.

As background to a man who played such an important part in the formation of modern sniping techniques, it seems appropriate to mention a little more about Burnham (1861–1947). He started his career at the age of 14 as a scout for the US Army during the Apache and Cheyenne wars. Then he went on to gold prospecting and cattle ranching and was at one point a deputy sheriff. Concluding that the Wild West was getting a little tame for his taste, he set off for Africa and signed up with the British South Africa Company as a scout. Eventually he became an officer in the British Army and a friend of Lord Baden Powell, founder of the Scouting Movement, to whom he taught bush craft. Besides serving as Chief of Scouts, Burnham was involved in numerous battles and escapades. Because of his superlative tracking skills the Africans called him 'He who sees in the dark'.

World War I

World War I (1914–18) was original in many ways. It was the first war, aside from the Napoleonic campaigns of France a hundred years earlier, in which the men of whole nations were conscripted to fight in national armies as opposed to having smaller professional armies doing the fighting for their country. World War I also came at a time when industrialization and the advance of technology including radio, aircraft, artillery and the machine gun allowed men to be killed in overwhelming numbers. Probably the issue which caused the greatest number of casualties was that the major protagonists were so evenly matched that this forced the conflict to descend to a war of attrition – a contest to establish which nation could stand the loss of its men and materiel the longest. In the battle between the French defending the city of Verdun, and the Germans attacking it, the French had 163,000 men killed and the Germans 143,000. The numbers of men seriously wounded were perhaps double this again.

Wars rarely go to plan and World War I was no exception. After the opening of hostilities in August 1914, the Germans made a quick advance westwards in accordance with their strategic goals, but they were halted by the French and the British (plus troops of their empires). This confrontation turned into the stalemate of trench warfare, where lines of trenches manned by infantry faced each other over some hundreds of metres of ground, churned by the heavy artillery that supported each side. For the first years of the war, defence held the advantage over offence, so there was little movement apart from periodic and often inconsequential offensives. In this environment, the sniper was king. Trench warfare was almost made for snipers because the two sides faced each other at ranges of a few dozen metres to hundreds of metres. The longer ranges presented too much of a challenge for the average issue rifle and infantryman to hit a human target, but not so to a new generation of marksmen armed with the best available rifles and firing the highest-quality ammunition. Sniping became something of an art form, and all the nations involved set up schools to turn out the best possible snipers.

The first snipers of World War I came from the ranks of the Germans, because they had been the only protagonists to set up sniper training schools before the war and issue their graduates with special rifles and optical scope sights. The Germans had the advantage of being one of the premier rifle manufacturers in Europe and the makers of the best lenses available at that time. Their telescopic sights had only 2 or 3x magnification to begin with, but even these magnifications were a massive advantage over standard 'iron sights'. And this magnification was sufficient for the capability of the rifles they were using, as an 800 or 1,000m (875 or 1,093yd) kill was a long-distance hit at that time. The tactics the Germans employed were largely the same as sniper teams apply today. They would deploy snipers in a team of two: a spotter to find the target and a shooter to make the kill. To begin with the British thought the long-range hits made by the German snipers to be lucky/unlucky shots (depending on your perspective) until they captured examples of the excellent Mauser sniper rifle equipped with telescopic sights. Snipers became a major drain on manpower and morale as, though they killed relatively few men compared with machine guns or artillery, men could not walk about freely or complete daily tasks without the constant thought that a sniper might be watching. This, and the fact that snipers also killed men employed in peaceable occupations, led to their being hated and shown no mercy when captured by either side. After a period of taking head shots from the British counter-snipers, the German snipers were issued with an armoured steel face mask to protect them. They were also issued with a piece of equipment called a 'hyposcope', which was a sort of periscope attached to a rifle, enabling it to be fired from below the level of the trench.

The first specialist snipers in British ranks were talented amateurs drawn from the wealthy chaps who had roamed about the empire big-game hunting. In fact, the first efforts at sniping were made with elephant rifles until service rifles were modified for the job. One notable character who was brought in to bat for the British was Major Hesketh Vernon Hesketh-Prichard, the noted hunter, explorer, writer, cricketer and soldier. He was shocked by the losses inflicted on British infantry when he realized that it was normal for a battalion to lose five men a day to snipers; one battalion had lost 18 in a day. He looked into the situation and found the German sniper armour would defeat the British .303in round, but the British armour, worn by machine-gunners and marksmen etc, was easily penetrated by the German Mauser.

Galvanized into action, and unable to gain financial support from the army or government, he collected and calibrated the handful of telescopic sights issued to the British and then called on celebrity friends across the world to send him hunting rifles and telescopic sights. Then he purchased more equipment with his own money and solicited donations from the public. Hex, as he was always called, investigated the way the Germans operated as snipers and discovered that they worked from a sort of raised parapet in their trench which was constructed in such a way that the sniper was not outlined nor visible while taking aim. The British, on the other hand, fired from the top of a normal straight trench and were clearly visible to the enemy. This was not a good idea at all, given that it was common amongst the snipers on either side to be able to make a head shot at several hundred metres. To counter this inequality, Hex developed the idea of using two sliding plates of armoured steel for a sniper or observer to peer through. As each only had a small aperture, and the two would be lined up at the viewer's target, the chances of the enemy being able to shoot back through them was massively reduced. The next thing Hex came up with was a device for finding enemy snipers. He had papier-mâché heads made, which were raised on sticks above the trench to attract sniper fire. These were even capable of holding a cigarette, to be puffed on by a soldier on the other end of a rubber tube for added realism. When the enemy sniper took a shot, it would obviously go through both sides of the head; the head was lowered and examined, then a trench periscope was used to recreate the relative alignment of the entry and exit holes. Thus the precise direction of the sniper was determined.

Hex eventually gained official support for his efforts and in August 1915 was given permission to set up a sniper school. By November of that year he was in such demand from so many units that he was made a general staff officer with the rank of captain. In August 1916, he founded the First Army School of Scouting, Observation and Sniping in the village of Linghem, France. In October of that year he was awarded the Military Cross, the citation of which read: 'For conspicuous gallantry and devotion to duty. He has instructed snipers in the trenches on many occasions, and in most dangerous circumstances, with great skill and determination. He has, directly and indirectly, inflicted enormous casualties on the enemy.' George Gray, a friend and himself a champion shooter, told Hex that he had reduced sniping casualties from five a day per battalion to 44 in three months in 60 battalions, and that by his reckoning this meant that Hesketh-Prichard had saved more than 3,500 lives. Hex published a summary of his work as *Sniping in France* (1920).

Using the excellent Lee-Enfield bolt-action .303in rifle, equipped with telescopic sights and precision ammunition, the British became at least a match for the German snipers by the end of 1917. The United States only became involved in World War I in April 1917, and began sending significant numbers of men to fight in Europe in the summer of 1918. As the war ended three months later, in the following November, there was little time for them to establish sniper units. Probably the most notable American sharpshooter of the conflict was Herman Davis (1888–1923) from Manila, Arkansas. He served as a private in the US 113th Infantry Regiment and became famous on 10 October 1918 for killing four German machine-gunners using a standard issue 1903 .30-06 Springfield Rifle and iron sights near the city of Verdun. The machine-gunners had pinned down Davis' platoon and his shooting allowed them to advance in safety. He attributed his shooting skill to hunting small game as a youngster in the Big Lake area. General Pershing later named Davis amongst the 100 greatest heroes of World War I and he received the Distinguished Service Cross (DSC), the *Croix de Guerre* with palm, the *Croix de Guerre* with gilt star and the *Médaille Militaire* awards from the American and French governments. After the war Davis returned home and died five years later during an operation related to tuberculosis, which he possibly contracted from exposure to poison gas.

Snipers of World War II

The world went to war again in 1939, but the technology was more advanced and the tactics were very different from the previous conflict. The German general Heinz Guderian had witnessed the British use of tanks to break through German lines in 1917 and 1918, and worked to formulate a new tactical theory for Hitler's Wehrmacht. He advocated using combined-arms formations of tanks, mobile infantry and light bombers (working as flying artillery) to sweep across a country too fast for a static defence to be mounted. This tactic was known to posterity as *Blitzkrieg* ('lightning war'), and when implemented in reality in 1939 and 1940 it was a stunning success, with Poland and much of Western Europe falling to Nazi occupation in short order. In June 1941, the Germans attempted to repeat their successes with an invasion of the Soviet Union, although that theatre was actually to prove their downfall.

Within this epic context, sniping continued and developed further. Some 19 years after the end of World War I, rifles were little different and the ranges at which they could kill were pretty much unchanged. What was new was the opportunity for sniper teams, freed from the confines of the trenches, to hunt down high-value targets and harass the advance of forces wherever they operated. The Russians, for example, had decided that sniping was a skill worth developing and the Soviet combat theorist Vassili Zaitsev contributed greatly to Soviet sniper doctrine.

During World War II the German Army had driven deep into the Soviet Union, and the Red Army was putting up a stiff resistance. One of the Soviet tactics was to deploy snipers to operate at will close up to the enemy and across the frontline. Zaitsev was one such sniper engaged in the defence of Stalingrad. (His exploits were so outstanding that a film was made about him starring Jude Law, called *Enemy At The Gate*.) In his first ten days at work, he killed 40 German soldiers. His final tally was 225. After the war he produced an autobiography and in it he recorded what were, in his opinion, the most important rules or tactics for a sniper to employ if he wanted to be effective and stay alive. Given his record, and the fact that he survived when so many did not, his opinions were valued. These are his 'top tips' for the sniper translated from the Russian:

1. Do not operate from a base-camp or fixed position. You must never remain long in one position as it might be observed and you will be killed by an enemy sniper.
2. Take care to gather all the intelligence you can.
3. Learn where, when, and how our own soldiers are being killed by enemy snipers. Then recce the area and work out how the enemy snipers are operating.
4. Work closely with the infantry soldiers deployed nearby. They will have a good idea of what is happening in their area of operation so they can provide intel for you and act as decoys or distractions.
5. You must always use a trench periscope for surveillance. It is never safe to expose yourself to gather intelligence.
6. You must always assume that apparent quiet hides danger.
7. Do your work slowly, cautiously and carefully. The successful sniper 'measures seven times and cuts once'.
8. When you absolutely must expose yourself, you should expose the barest minimum possible; that minimum should be totally camouflaged, and remain still while in view.
9. Much of your time you should lie like a stone and merely observe. Your aim is to be completely invisible to the enemy, even to the trained observer.
10. Your war is one of nerves, concentration and endurance.
11. Create distractions for the enemy, exasperate him with diversionary movements and finally exhaust his patience and ability to concentrate.
12. Set up a dummy close to your own position. This dummy is designed to draw the attention and fire of an enemy sniper so that he will reveal his position.
13. As an alternative, allow the enemy sniper to become used to your dummy until he becomes careless about it. Then take its place and get the enemy sniper in your crosshairs.
14. You must not fire until you are absolutely certain of a kill, as if you fire without a clear target you will reveal your own position.
15. Everything depends upon maximum intelligence, meticulous preparation, careful attention to detail and endless patience.
16. Impatience is death.

Zaitsev found that military organizations have difficulty replacing experienced NCOs and field officers during times of war. He also reasoned that the more expensive and less rugged sniper rifles could match the cost-effectiveness of a cheaper infantry rifle given good personnel selection, training and adherence to doctrine. The result of Zaitsev's research was twofold. The Russians decided to put a lot of effort into producing large numbers of snipers, and they realized that sniping was not all about the extreme long-distance shot – it was as much about being in the right place at the right time and aiming at the right man.

The Russians were short on weapons, manufacturing capacity and money, but what they did have was a vast supply of compliant manpower. Given this huge human resource, and a limited supply of weapons, the Soviet leaders hit on the idea of selecting the best possible female recruits as snipers and training them up so they could be used to harass the invading German forces. The result of the drive to select female potential snipers was the recruiting of 1,500 young women from amongst party members of proven loyalty and unquestioning idealism. They were all in superb physical condition and had excellent eyesight. What the leadership thought they might lack, however, were the skills in outdoor living and hunting expected of their male counterparts; but these could be taught, while hawk-like eyesight and determination could not. (As a side note here, the male sniper recruits who volunteered were usually from a hunting-type background and so were already familiar with stalking and shooting. They were quickly sent to the front and even more quickly became either very good or very dead. Nevertheless, the best female snipers were as good as some of their male counterparts.)

The female recruits eventually arrived at the Central Female Sniper Academy in the village of Veshnyaki near Moscow. For the first eight months they were rigorously trained in tactics, camouflage, field craft, marksmanship and physical conditioning, while their belief in the Communist Party line was continually reinforced. In combat, the female snipers achieved long-distance kills similar in difficulty to those performed by the men. They also displayed the same cunning and patience in stalking their targets and they endured the same conditions, overcame the same perils and used the same weapons.

The women served as equals, but they functioned with a noticeably stronger communist fanaticism that Soviet propagandists recognized and exploited. In an interview, a female sniper remarked, 'I always carried two grenades, one for Fritz and one for me.' The women had no illusions as to their treatment if captured and their attitude showed well in the defence of Kiev.

The entire Soviet Army achieved 14,568 confirmed sniper kills during World War II, which was the equivalent of immolating a full German division. Doubtless the real score was much higher despite the propaganda. The top Russian female snipers kept score of their victories and competed amongst themselves for the most kills. Two well known rivals were Nina Lobkovskaya with 308 confirmed kills and Lyudmila Pavlichenko at 309, both Russian citizens. (Pavlichenko was credited with being the top female Red Army sniper in World War II, with the highest kill count of enemy soldiers, and amongst them 36 confirmed enemy snipers, one having 500 kills documented in the journal discovered on his body.) Propaganda is vitally important in any state, let alone a starving, repressive police state, so the Department for Agitation and Propaganda, as it was endearingly named, made the best possible use of the pretty snipers, and they appeared regularly in newsreels shown to the masses in cinemas across the empire, often with killing scenes re-enacted for the benefit of the cameras. The Communist Party turned the lady snipers into rock stars and Pavlichenko received the Gold Star of the Hero of the Soviet Union. She even toured the United States and Canada on a goodwill trip meeting Franklin Roosevelt at the White House. Of course the Americans and the Soviets were on the same side in those happy days. The top Russian male sniper was Mikhail Surkov of the 4th Rifle Division with 702 confirmed kills over the course of the Axis invasion.

The two sniper rifles employed by the Soviet Union from the beginning of World War II were the Mosin-Nagant PE and the semi-automatic Tokarev SVT-40. The Mosin-Nagant PE had been in use since well before the war and was based on the standard 1891/30 bolt-action infantry rifle chambered for the 7.62 x 54R cartridge, but the parts were hand selected for quality. A 4x telescopic sight was added; it was originally a copy of a German scope, but later the Russians developed a design that was, in theory, easier to mass produce.

The Tokarev SVT-40 was intended to replace the Mosin-Nagant at the beginning of the war, but it proved to be a disaster. It was difficult to manufacture, it kept jamming, the muzzle flash was startling and it was often inaccurate due to poor engineering. It was quickly scrapped and the designers and factory managers presumably shipped off to Siberia. To replace the SVT-40 the Russians decided to return to what they knew with an upgraded version of the Mosin-Nagant PE, which they designated the PU. It was made easier to mass produce and the improved, simple scope from the SVT-40 was adopted. It has proved to be the longest-serving sniper rifle in the world and lasted in frontline service with the Russians until it was replaced in 1962 by the semi-automatic Dragunov. Of which more later.

A potentially very important fact became apparent from analysis of the German advance into Russia and their subsequent retreat. During the German advance, the Russian snipers had the advantage and when the Germans were retreating the Germans had the advantage; with no change in tactics, personnel or weapons. This suggests very strongly that withdrawal favours the sniper when he, or she, is tasked with harassing an advancing conventional army. At least, to be specific, withdrawal favoured the sniper statistically at that time with those weapons, those tactics and those conditions.

The Germans actually began World War II relatively unprepared for a sniping conflict. There was little investment in sniper training following World War I; individual infantrymen could attain marksman status using a variety of Gewehr 98 rifles and hunting weapons, fitted with telescopic sights, and were hence repurposed as snipers. It was the experience of encountering Soviet snipers from June 1941 onwards that effectively changed army policy. Battalion commanders were ordered to select the best shots from their men to form separate sniper units, typically around 22 soldiers strong, who would serve among the frontline companies. As the war progressed, a number of German sniper schools were established, to teach the field craft and technical skills that came with the job. Note, however, that the emphasis lay not so much on shots at extreme range, but precision shooting at between 400 and 800m (437 and 874yd). The German snipers were also trained to be more selective in whom they shot, when compared to their Soviet counterparts, as sniper historian Martin Pegler has noted:

Generally it was not so much the number of kills the German snipers made that made them invaluable, but the quality of hits. Russian troops were not good at advancing without leadership, and depriving them of control would usually lead to an immediate cessation of attack. The sniper's ability to target commanders or senior NCOs made this sort of tactic invaluable – one German sniper noted that he prevented an enemy advance by shooting eight Russian commanders in one day.

– Martin Pegler, *Out of Nowhere* (Oxford, Osprey Publishing, 2011), p.151

Such intelligent targeting policy was sometimes augmented by some rather more ruthless tactics. Pegler goes on to explain how the German sniper Franz Kramer would, when facing a Soviet assault, shoot Soviet soldiers in the stomach, to demoralize the attacking force with the shrieks of the wounded.

German sniping against the Western Allies was no less brutal, but by the battle for Normandy in June 1944 the Germans had developed sniping into a fine art, especially in terms of camouflage and field craft. Individual snipers would secrete themselves in trees, barns, ditches and buildings, and from there took a severe, selective toll on the advancing US and British troops. Often the only Allied response was to open up with massive retaliatory firepower as soon as a muzzle flash was identified, and many German snipers came to an end in a hail of machine-gun and mortar fire.

The German learning curve was as much technical as tactical when it came to sniping. Soviet snipers killed or caught with semi-automatic rifles such as the SVT-40 prompted development of similar German weapons, such as the Gew 41 and Gew 43. Neither were actually conceived as pure sniper weapons, but once fitted with telescopic sights they were serviceable precision weapons out to about 600m (656yd). By far the majority of sniping, however, was performed with classic bolt-action weapons. Many of these were Gew 98 rifles, albeit with refurbished actions to ensure more consistent accuracy. Eventually some dedicated sniper weapons were produced and distributed to the frontline, mainly the Turret Mount K98k, an adaptation of the standard infantry rifle but with a much-improved scope-mounting facility. The weapon was not perfect (few World War II sniper rifles were), but it performed deadly service in the hands of German snipers on all fronts. The most significant hindrance to the snipers doing their duties properly was actually manufacture and distribution. With so many weapons being destroyed or lost on the Eastern Front, the output of sniper rifles always lagged well behind demand, especially as the German war machine attempted to keep up with multiple competing demands for men and materiel.

By 1944 the Allies had also caught up with the art of sniping. The Americans began their war with almost no practical sense of sniping nor an effective sniper course. The US Army did have a reasonable weapon for long-range shooting – the .30-calibre M73B1 with a 2.5x Weaver 330 telescopic sight, actually a specialized sniper version of the M1903 Springfield rifle. The Marine Corps used the same weapon, but fitted with a more powerful Unertl scope, and even adopted the Winchester Model 70 commercial rifle using the same scope. From 1944 the US troops also had accurized .30-calibre M1 Garand rifles fitted with Lyman Alaskan 2.2x scopes, which gave them a semi-auto sniper facility. The Garand had already provided US forces with a firepower advantage in conventional infantry engagements, and its build quality was sufficient to deliver dependable sniping work over medium ranges. Furthermore, if the sniper found himself confronted with enemy forces at close range, his Garand could generate decent suppressive firepower, something that couldn't be said of a bolt-action rifle.

The demands on US snipers varied according to the theatre in which they found themselves deployed. In the Pacific, sniping tended to occur at relatively short ranges on account of the nature of the jungle terrain. Nevertheless, the US Marine Corps in particular prided itself on the field craft and tallies of its snipers. The Corps also produced some outstanding shots. One private, David Cass, demonstrated his consistent skill by shooting the entire crew of a Japanese machine gun at a distance of 1,143m (1,250yd) during the battle of Okinawa in April–June 1945.

In return, the Japanese also fielded large numbers of snipers, who tended to hide out in the dense jungle tree tops, sometimes even lashing themselves to the tree trunks to ensure their stability and their commitment. As time went on the US soldiers grew wise to the enemy hiding places, and would hose down likely trees with automatic fire as a matter of course. Consequently, the losses of Japanese snipers to US (and British) equivalents were always disproportionately weighted in the Allies' favour, but the presence of Japanese snipers remained significant and mentally wearing.

During 1943–44, the US forces expanded their theatre of operations to include North Africa, Italy and Western Europe. Sniping also grew increasingly structured. In a 1944 US sniper manual, two types of sniper role were defined. The first was 'observer-sniper' teams, two-man units operating from fixed camouflaged posts and observing the ground in front of them for both intelligence and targets. Military historian Gordon Rottman here explains something of the operational procedure for the observer-sniper team:

> To avoid fatigue observation duty alternated every 15 to 20 minutes. Where possible range cards were drawn up for each post, showing landmarks and distances to designated points. These would allow for quick direction of the shooter to targets, and accurate shooting on known data. Concealment and patience were primary requirements for the observer-sniper team, with rifle barrels not to protrude beyond cover and smoking strictly forbidden.
>
> – Gordon Rottman, *World War II Infantry Tactics: Squad and Platoon* (Oxford, Osprey Publishing, 2004) p.54

The two-man observer-sniper remains a classic sniper team structure to this day, although in modern US terminology the team is described as spotter and sniper. During World War II this small unit was in contrast to what the US sniper manual described as the 'mobile sniper', a lone individual sent out to hunt enemy personnel aggressively. The manual describes his role as follows, plus cautions about the demands placed upon his marksmanship:

> The mobile sniper acts alone, moves about frequently, and covers a large but not necessarily fixed area. He may be used to infiltrate enemy lines and seek out and destroy mobile targets along enemy routes of supply and communication. It is essential that the mobile sniper hit his target with the first round fired. If the sniper is forced to fire

several times, he discloses his position and also gives the enemy time to escape. Therefore, although the mobile sniper must be an expert shot at all ranges, he must be trained to stalk his target until he is close enough to insure that it will be eliminated with his first shot.

– Quoted in Rottman, p.54

The mobile sniper described here performed a more focused role than the observer-sniper team, actively inflicting attrition and disorder upon the enemy's supply lines and communications. Yet regardless of the type of sniper, the unnerving effect upon the enemy could be the same. One of their principal challenges was to serve in a counter-sniper role – talented and experienced German snipers were taking a worrying daily toll of US soldiers. One German sniper, Karl Krauss, noted that he was able to hold up an entire column of US vehicles for hours during the Italian campaign by firing just five rounds, forcing the American soldiers to take cover and move with extreme caution. Furthermore, many German snipers found their US equivalents somewhat amateurish. Krauss himself noted that 'Sometimes we came under fire from the GI snipers, but they were often not good shots, and their hidings were usually very easy to find. They were not trained well, I think, and once we captured two. They had good rifles with small telescopes but they did not know much of camouflage and they fired too many shots from their hiding place, so we found them' (Krauss, quoted in Pegler p.187). Krauss' judgement that poor training was to blame for the soldiers' capture is probably accurate. Although the instruction given to snipers did improve throughout the war, it remained rather haphazard, and the training in field craft and 'scouting' depended very much on the experience of the instructor. In the 1944 manual *Scouting, Patrolling, and Sniping*, the following passage outlined the requirements of sniper training:

Within each platoon, several men will be given sniper training. These men will be selected from among the most proficient marksmen in the unit and will be given training in scouting and camouflage and in the use of the sniper's rifle. Snipers, in addition to being expert shots, must be trained to estimate ranges accurately, to select advantageous firing positions, to move silently through difficult terrain, and to be proficient in the use of maps, aerial photographs, and the compass. Also, they must be physically agile and hardened and able to sustain themselves for long periods of detachment from their unit. One of the men undergoing this special training will be designated to carry the sniper's rifle, but the platoon leader may, upon occasion, designate other men to act as snipers, employing carbines or rifles which do not have the telescopic sight.

– War Department, *Scouting, Patrolling, and Sniping* (1944) p.173

The manual then goes on to list how these specific skills are to be developed, but the last sentence of the paragraph above is interesting. It suggests that good shots from within the regular infantry could find themselves suddenly appointed as snipers, using only their standard weapons and iron sights. Such might account for Krauss' observations above about the quality of the US snipers.

Despite the German viewpoint on the American shooters, the US soldiers inflicted significant casualties upon the enemy. The greener US snipers had to learn fast if they were to take their battle to the enemy, but learn they did. Sergeant John Fulcher of the US Army was one half of an observer-sniper team deployed to Salerno in Italy in September 1943. Here he describes what occurred one morning, as he and his spotter set up observation of a track and saw a German company come marching into view:

I looked at my partner. He had his rifle scope trained on them. He looked back at me. He shook his head. A whole company. I nodded at my partner. Let's take them. They're green. Even if they organized an assault, we could be off the ridge before they got halfway across the field to us. As cool as I could be, I cross-haired the officer and shot him through the belly. He was dead by the time I brought my rifle down out of recoil and picked him up again in my scope. His legs were drumming on the road, but he was dead.

– John Fulcher, quoted in Pegler, p.186

The rest of the company naturally scattered when their officer fell, and retreated back down the track. However, another pair of snipers were actually waiting further down the road, and the German company took more casualties until they finally made their escape.

Just as the US forces had to adjust to a steep learning curve in terms of sniping, so too did the British Army. The British had discovered the value of sniping during World War I, but in 1939 there were few rifles designed for sniping and few men trained to use them properly. The problem of weaponry was solved initially by converting nearly 1,500 of the new No. 4 Lee-Enfield infantry rifles to sniper standards, by the addition of a wooden cheek rest on the stock plus the mount for a No. 32 telescopic sight. The weapon was subsequently known as the Enfield No. 4(T), and it was soon in the hands of British snipers in North Africa. (The sporting gun manufacturer Holland & Holland were later contracted to convert 12,100 of the rifles to the No. 4(T) standard.) There it proved to be a decent long-range weapon, reaching out confidently to targets beyond 400m (438yd), even with the standard .303in ammunition.

The evolution of British sniper training mirrored the experience of the US forces. Training during the early years of the war was uneven, and based heavily upon theory and experience gained during World War I. Gradually, however, sniper schools were established at various locations within Britain and even in Britain's overseas possessions, and began to turn out a new generation of battlefield hunters.

Much about the British sniping system remained improvised. In the absence of specialized sniper clothing, for example, camouflaged paratrooper Denison smocks were often worn, the concealment refined with the addition of vegetation and scraps of fabric. In the battle for Normandy in 1944, the British snipers had to settle into the job quickly if they were to inflict punishment on the enemy, while also staying alive under the threat of talented and experienced German snipers. Sniper sergeant Harry Furness later remembered that 'I haven't the words to really describe thesheer horror of fighting as a sniper in Normandy. That is why I never went back for fifty years, you don't go to places that were nightmares' (quoted in Pegler, p.205). The infamous *bocage* terrain of Normandy meant that snipers might find themselves engaging enemy troops at close quarters as much as at long range, and casualties were appallingly high on both sides. Snipers also found themselves in terrifying urban battles, as Furness goes on to recount:

As our rifle companies fought house-to-house, and room-to-room in close combat, all of us snipers were kept hard at it in support. This type of fighting is horrendous ... as the retreating Germans fought us every inch of the way. With their infantry inside every house and their tanks and SPs [self-propelled guns] roaming the streets, they fired high explosive shells into every building we occupied, so often our snipers were shooting from burning buildings .... there were so many places which concealed German snipers that ... it was difficult to judge where the shots were coming from.

– Quoted in Pegler, p.201

This quotation illustrates perfectly how we should not imagine all snipers operating in a quiet 'bubble' of concentration, away from the general violence and disorder of the battlefield. The snipers of World War II had to fight the same war as everyone else. The war ended, however, with a cemented understanding of sniper operations and how snipers could contribute to modern warfare, lessons that would be carried forward into post-war conflicts.

Vietnam & the .50-cal

Although World War II ended in September 1945, fighting continued in new conflicts throughout the world. South-East Asia was a case in point. From 1945 to 1954, Indochina (consisting of Vietnam, Cambodia and Laos) was plunged into a devastating war between communist insurgents and regular forces, led by the infamous Ho Chi Minh, and French colonial troops attempting to take possession of the region again following its wartime occupation by the Japanese. The conflict ended with a French defeat at Dien Bien Phu and a Vietnam divided into a communist North and a US-backed South. This situation was not to last, and a growing communist insurgency in South Vietnam led to direct US involvement in the Vietnam War between 1963 and 1973, and a communist take-over of the South in 1975.

The Vietnam War was an extraordinary conflict, with the combat ranging from low-level US search-and-destroy missions to major conventional battles in Vietnam's jungles and cities. The sniper was by this time central to US tactical thinking. Snipers provided the means to go deep into enemy territory and take out Viet Cong (VC) and North Vietnamese Army (NVA) personnel in regions where they thought they were safe. Much of Vietnam is covered in thick jungle, but there are areas of lighter cover and, of course, roads and other clearings. The effect of this terrain on the task of a sniper was that very often he could get close to his target without being detected.

There were also many brave, and some skilled, men and women working as snipers amongst the Vietnamese insurgents both during the French occupation and the US war. Presumably because such individuals were mostly uneducated, there are very few records of their exploits available. What has been well recorded is the experiences of the occupying forces who were their targets. In the following transcript, a US soldier records his experience of a Viet Cong sniper taking shots on Thanksgiving Day:

> Monsoon season and rain and mud everywhere, but chance to relax. We wrote letters home and just shot the bull without our flak vest or weapons at the ready. Anyway kinda misty and mild fog and then heard sound of the thump of rotors of choppers coming in and to our surprise it was this hot chow for this holiday with turkey with all the fixins. Anyway as we lined up to indulge in this special occasion with our mouths watering and thinking of our families back home, and getting into the holiday spirit, we laughed and goofed around and forgot our everyday worries. Think I even saw you and [Lieutenant] Pontuck laugh some. Anyway some pitiful V.C. thought he would harass us with some shots from the side of the hill maybe 200 yards away. Wasn't doing any damage, just being pain in the ass with us 'cause it was raining and chow was steaming and waiting for us to devour. As we crawled through the chow line for our turkey and cranberry relish with drops of rain trying to make our potatoes soggy, we were all so happy even though when eating our hands were muddy and our clothes soaking wet.

The Vietnam conflict spawned many notable snipers amongst the US ranks. Chuck Mawhinney, for example, was one of the USMC's top snipers in Vietnam, but kept quiet about his exploits – not even his wife knew for many years. Word got out when a friend wrote a book highlighting what he had done. The book, *Dear Mom: A Sniper's Vietnam*, publicized Mawhinney's record of 103 confirmed kills in Vietnam, and another 213 more unconfirmed. Mawhinney left Vietnam in 1969, after 16 months as a sniper. After a short stint as a rifle instructor at Camp Pendleton, Mawhinney then also left the Marines and returned home to rural Oregon. 'I just did what I was trained to do,' he told *The Standard*. 'I was in-country a long time in a very hot area. I didn't do anything special.'

Similarly, Carlos Hathcock was a USMC sniper with a service record of 93 confirmed kills. Hathcock's record and the extraordinary details of the missions he undertook have made him a legend in the Marine Corps. During the Vietnam War, kills had to be confirmed by an acting third party, who had to be an officer, besides the sniper's spotter. Snipers often did not have an acting third-party present, making confirmation difficult, especially if the target was behind enemy lines, as was usually the case. Hathcock is ranked fourth, behind USMC snipers Eric R. England and Chuck Mawhinney and US Army sniper Adelbert Waldron, on the list of most confirmed kills for an American sniper. He probably killed a great many more than his official record, however. The NVA put a bounty of $30,000 on Hathcock's head as he had killed so many of them. They called him 'White Feather' owing to the white feather he wore in his hat. When a platoon of Vietnamese snipers were sent to hunt him down, many Marines in the same area took to wearing white feathers in their hats in his support. Once Hathcock shot an enemy sniper down the length of the sniper scope and hit him in the eye after seeing the sun shine off the glass of the lens. Hathcock concluded that, for his bullet to run down the scope, the enemy sniper must have been lined up on him and ready to shoot but he got the shot in first.

Hathcock also volunteered for a mission to kill an NVA general just before the end of his deployment. He crawled 1,370m (1,500yd) over four days to reach the shooting position, being almost stepped on by the enemy and very nearly bitten by a bamboo viper on the way. The general came out of his tent and Hathcock hit him in the chest with his shot, killing him instantly. Then he had to crawl all the way back again while the enemy began a vengeful search.

On his operations, Hathcock generally used the standard sniper rifle (Winchester Model 70 30-06) with an 8x scope at moderate ranges, but he did experiment with the Browning .50in machine gun for longer-range shooting. At the time, the metallurgical technology did not exist to build a light .50in sniper rifle, so to achieve long- range kills with this heavy, and potentially very accurate, round it was necessary to use the Browning M2HB machine gun to fire it. Hathcock mounted a 10x Unertl scope on the Browning, using a bracket of his own design, and 'single tapped' the rounds with the trigger (not a difficult procedure owing to the Browning's slow cyclic rate of fire). Using this weapon Hathcock made a number of kills at ranges in excess of 914m (1,000yd), including his record for the longest confirmed kill at 2,089m (2,286yd). This record was not beaten until 2002 during the war in Afghanistan, where snipers are equipped with far superior rifles. Hathcock said it was the hunt, not the killing, that gave him the kick.

It should be noted that although most US snipers in Vietnam relied on bolt-action weapons, the US Army also wanted an accurate semi-automatic sniper rifle, with the ability to take a quick second shot. The 7.62mm M14 model was selected because it was gas operated, accurate and reliable with a 20-round magazine. For sniper use, each unit was built from the most precise parts to 'target grade'. This rifle was fitted with a 3–9x adjustable ranging telescope and match grade target ammunition was supplied for each rifle. This combination was designated the M21 and remained the army's sniper rifle from 1969 until 1988, when it was replaced by the M24 Sniper Weapon System (SWS).

Recent Wars & Shooting Records

Many of the lessons learned in Vietnam about modern sniping went to inform sniper training in the post-Vietnam era. Roughly a century after the sniper concept had begun to emerge in military forces, snipers had become a fundamental part of the armed forces of the late 20th century, and with new ultra-sophisticated weapons they took the art of sniping to even greater levels of ruthless efficiency. Such is evident in the very recent wars in Afghanistan and Iraq, fought by US-dominated coalition forces since the September 2011 terrorist attacks on the United States.

The wars in Iraq and Afghanistan have presented snipers with some similar, and some contrasting, challenges. Both war zones have involved the stress of telling friend from foe, where an insurgent enemy is often visually indistinguishable from the civilian population, and the near-constant threat of sudden ambush. Both have also witnessed the scourge of improvised explosive devices (IEDs), explosive booby traps of varying ingenuity that have been the primary source of coalition casualties – catching a bomb-maker in the act of setting his device is a dream for most US, British or Allied snipers.

By far the biggest contrast has been in terms of terrain. In Iraq, much of the conflict has been urban in nature, with snipers operating at the closer ranges and amidst the visual confusion of a town or city. In Afghanistan, coalition snipers have operated in very mountainous wilderness, often in support of major conventional operations against Taliban and al-Qaeda forces in high-altitude retreats. For example, in March 2002 the US forces launched Operation *Anaconda*, in which 1,700 US troops and 1,000 Afghan military, together with other coalition units, attempted to contact and destroy a group of insurgents estimated to number between 200 and 1,000 in the Shah-i-Kot Valley. This was the first operation in which a large number of US conventional troops, as opposed to special forces, were pitted against insurgents in direct combat. The insurgents established in the mountainous area fired mortars and rockets from entrenched positions, while Afghan Taliban commander Maulavi Saifur Rehman Mansoor led further Taliban reinforcements to the scene.

Canadian sniper Corporal Rob Furlong of the 3rd Battalion, Princess Patricia's Canadian Light Infantry, with his sniper team (two corporals and three master corporals), was part of Operation *Anaconda*. During that operation he set what was then the world record distance for a sniper kill at 2,430m (2,658yd). Armed with a .50in Tac-50 Rifle loaded with A-Max low-drag ammunition, he sighted a three-man insurgent weapons team setting up their position anddecided to take the shot. His first round missed, his second round hit the target in the knapsack on his back and alerted him, but the third round followed quickly and hit centre mass. The flight time at this range, for a bullet fired from this weapon, is around three seconds.

The conflict in Afghanistan has produced other record shots. British sniper Corporal Craig Harrison of the Household Cavalry was on a vehicle-mounted patrol in Helmand Province when he saw two insurgents advance across a compound carrying a machine gun, and open up on his commander's vehicle some way ahead of him. From his own vehicle, admittedly in what he described as 'perfect' conditions, he took aim at the gunner with his British-built 8.59mm L115A3 Long Range Rifle. His first bullet struck the gunner in the stomach and he went down. The gunner's number two then tried to fire the machine gun, but he was hit in the side of the chest and died instantly. This fairly straightforward 'war story' becomes extraordinary when you consider that the range was 2,475m (2,707yd), a distance even greater than that established by Corporal Furlong. Harrison killed a further 12 insurgents and wounded seven others, and survived a bullet which went through his helmet and a roadside bomb that broke both his arms. He quickly returned to duty with his shooting unaffected.

It should be noted that the sniping in Afghanistan has not only been one way. By April 2010, one particular Taliban sniper had been stalking British troops – specifically the 3rd Battalion The Rifles – in the Helmand Province town of Sangin for five months, and had killed seven of them. One victim was even a British sniper sent out on a counter-sniping mission against this insurgent. But this may not be the total kill count of this insurgent sniper – 53 British servicemen have been killed in the area over the last 12 months (at the time of writing, November 2011), which that is about 12 times the average casualty rate for a similar area. Clearly not all of these are victims of this particular sniper, but some are likely to be. The British think he has been trained in either Pakistan or Iran, as he seems to have a good grasp of the tactics and skills required to perform his task and he knows things an amateur would not – such as how to watch and establish patterns of enemy movement.

Iraq has also spawned its fair share of insurgent snipers. In 2006, for example, sniper Muhammad Awwad Ahmad was operating in Fallujah, Iraq, when he shot and killed Lance Corporal Michael Glover and then Captain John McKenna as he tried to pull Glover to safety. Ahmad was captured by the US military and handed over to the Islamic civil authorities and imprisoned. Now that US troops have largely left Iraq, he has been released from prison after serving five years.

However, although the toll inflicted by insurgent snipers in Iraq has been significant, the balance of kills is by far unequally weighted in favour of coalition snipers, some of whom have performed shots demonstrating unusual levels of skill. In June 2003, Royal Marine sniper Corporal Matt Hughes was ordered to eliminate an Iraqi sniper who was holding up the British advance in southern Iraq. The enemy sniper thought he was secure as he was hidden from the main unit's view, and he fired repeatedly. Yet from his vantage point Hughes could see the head and chest of his target, still wearing the Iraqi Army green uniform. The range of 860m (941yd) was not a problem for his 7.62mm L96 sniper rifle, but there was a gale blowing from left to right. Hughes' spotter gauged the offset required by studying the movement of heat haze and dust across the arid desert landscape and Hughes took the shot. The bullet curved 17m (56ft) sideways in the air and hit the target in the chest, killing him instantly. Another Royal Marine sniper killed a second Iraqi, who was standing next to the insurgent sniper, at exactly the same moment.

On the US side, there were equal talents at work. Staff Sergeant Jim Gilliland, 28, was leader of 'Shadow' Sniper Team and stationed in Ramadi, Iraq during January 2006. Over the previous five months his team of 10 men had killed almost 200 insurgents between them. On 1 January he was looking through the telescopic sight of his M24 rifle at an insurgent who he had just seen kill an American soldier. His target stood, looking relaxed, in the fourth floor bay window of a hospital, holding a Russian Dragunov sniper rifle to his chest. Gilliland allowed for wind and the drop of the bullet and squeezed the trigger. The single shot hit the insurgent in the chest from a range of 1,250m (1,368yd) and dropped him where he stood. 'I believe that is the longest kill in Iraq with a 7.62,' said Gilliland, who hunted squirrels as a boy in Alabama before working his way up through deer to people. Later in the day Gilliland found out that the soldier he saw killed by the insurgent sniper was his buddy Staff Sergeant Jason Benford.

Alongside regular army and marine snipers such as these, special forces snipers are also heavily employed in Afghanistan and other war zones. Between them, these snipers make the insurgent's life a perilous one, not least because of the weaponry they carry, which will be the subject of our next chapter.

CHAPTER 2
THE LONG-RANGE WEAPON

What are the main factors that determine the accuracy of a shot? Is it the rifle or the shooter? Could a super-marksman achieve great hits with an old Brown Bess musket? What about a novice with the latest sniping rifle?

In reality, there are three separate groups of factors which affect the accuracy of any sniping shot. These are:

- Conditions – wind, humidity etc.
- Shooter – his training, technique and ability.
- Rifle – the capability of the rifle itself.

Over much of the remainder of this book, we will be looking at each of these factors in depth. Here, however, we will look mainly at the issues affecting the inbuilt accuracy of the weapon itself. For while it could be said that 'A poor workman blames his tools', there is a limit to what a sniper can do with an inaccurate rifle.

Accuracy

Before we examine the factors affecting accuracy, we first need to understand what accuracy really is, how it is measured and what its limits are. Unfortunately for the student, ballistics – the study of projectiles in motion – is a scientifically complex subject, and the behaviour of a rifle bullet in flight is no simple matter to consider. If a rifle is fastened rigidly into a vice, as armourers and engineers do when testing them, and several rounds are fired, the bullets will hit the target at random over a small, roughly circular area, rather than go through the same hole. The tiny variations in the bullet and the barrel will send each bullet high or low, left or right, just a little. Minor variations in the type and quality of propellant will affect bullet flight because of muzzle velocity effects against rotational inertia, as will be seen shortly. And as the rifle barrel warms, the propellant will burn faster. The point here is that the rifle will fire a 'group', rather than putting all the bullets through one hole in the target, even when it is clamped to a bench.

The smaller the group a rifle makes when it is clamped into position and fired the more accurate it is. What is a good group and what is required of a sniping rifle? For the moment, we need to understand the idea of measuring the variation in what is called 'fall of shot' – where the bullets actually land – to compare the accuracy of two rifles. To keep it simple, the pattern made by the fall of shot is circumscribed by a circle; the smaller the circle, the more accurate the rifle, and the tiny offset between the edge of this circle and the centre of the circle, where the barrel is pointing, is represented as an angle. That angle, often referred to as Minute of Angle (MOA) – a minute is 1/60th of a degree – gives the accuracy of the rifle/ammunition combination. The shooter is taught to place his hits within as small a group as possible on the target, then the sights are adjusted so as to bring the centre of that group of shots onto the place where he is aiming.

A layman might be excused for thinking that a bullet which is 2.5cm (1in) away from the aiming point at 100m (91yd), roughly 1 MOA, is going to be 25cm (10in) away at 1,000m (1,093yd). Actually it is not. The bullet's velocity and its rotation slow in flight, its temperature changes and so on. For this reason, when an army is specifying the required accuracy for a sniping rifle it does not just say 'Must fire a group no larger than X inches at X hundred yards'. What it does is specify the spread of shot at various ranges up to the maximum requirement, so it knows the rifle's capabilityfor accuracy at, say, 200m (219yd) as well at 1,200m(1,321yd).

Today the modern ordnance department no longer expresses accuracy in terms of the size of a group shot at a certain range, as it did when things were less technical. Now things have to be precise: We have two terms; the first is MOA which we have seen above and the second is MRAD, which is a measure or arc based on the relationship between the radius of a circle and the distance between two points on the circumference of that circle. Rather than looking further into the maths it is sufficient for our purposes here to know that you can convert degrees to radians by the following simple sum: Degrees x 180 over π = radians. A 1 MOA for a five-shot group – measured as the centre-to-centre distance between the outside bullet holes – translates into a 69 per cent probability of putting your shot within a circle of 23.3cm (0.9in) at a range of 800m (875yd). This is actually the standard generally required to hit a human torso at that range.

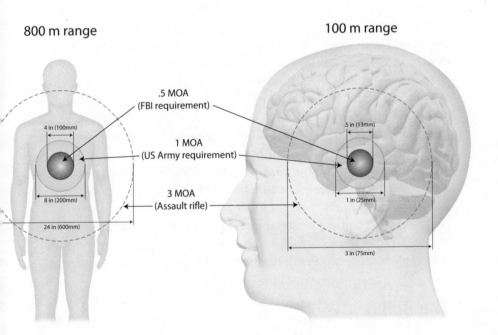

Comparison of 0.5, 1 and 3 MOA extreme spread levels against a human torso at 800m (875yd) (left) and a human head at 100m (109yd) (right).

Note that there are different accuracy requirements for different shooting jobs. A police sniper generally is shooting from less than 200m (219yd). An infantry soldier in close country rarely gets to hit anyone much more than 100m (91yd) away and is not concerned about a clinical kill. A designated marksman (DM) is added to a patrol of infantry in open country to extend the range of their assault rifles and he might be expected to hit a target out to 500–800m (547–874yd). A true army sniper, however, is trying to hit a target at the maximum possible range available, as this opens up the number of targets he can take on. Special forces snipers are somewhere amongst these requirements, as they may be given a short-range target which is vital or a long-range task where the circumstances, perhaps climate, make the operation arduous or otherwise difficult. Nevertheless, a true sniper wants to be sure of a kill at the greatest range possible and will expect that to be more than a mile.

The sniper rifle needs to support all these demands. A military-issue assault rifle is usually capable of 3–6 MOA accuracy because this modest requirement allows it to be light and rugged. A standard-issue military sniper rifle is typically capable of better than 1 MOA accuracy out to extended range, so it will fire a heavy round and require gentle handling to protect the optics. A police sniper rifle should be capable of 0.25–1.5 MOA accuracy to make an instant kill, but is only required to achieve that over short range so it can be a lighter weapon. For comparison, a competition target rifle may be capable of accuracy 0.15–0.3 MOA, but is designed to work on a range where it can be handled gently, and where there is no dirt and no hurry. Special Operations Forces will select the rifle for the task in hand: a heavy round for long range, a lighter rifle and round if it has to be carried a long way and perhaps a very light or 9mm sub-sonic weapon if it has to be hidden or the weapon silenced.

The Bullet

A bullet is the projectile fired by a rifle, pistol or other firearm. It normally achieves its aim of damaging its target by the kinetic energy or blow which it delivers to the target through making contact at a high speed; often in excess of 2,413km/h (1,500mph).

This contact results in either a hole in the target through which blood escapes or, because of the shock wave running laterally and conically ahead of the bullet through the target, a large chunk of flesh being ripped out. There are some bullets which are designed to expand on impact, or even carry an explosive charge, to do more damage but these are rarely used by a sniper as these modifications of weight and bullet tip tend to reduce the extreme accuracy which he requires.

A sniper generally needs to be able to hit a human sized target, and do serious damage, out to the maximum possible range (currently around 2.4km/1.5 miles). To do this he needs a bullet that will go where it is aimed for as far as that distance and which is soft enough to deform and cause maximum damage to the target, while being robust enough to deliver some penetration. That is achieved by making the bullet dense-cored, hard-jacketed, streamlined and boat-tailed to reduce the drag coefficient. But accuracy and killing power are not the only requirements or considerations. If the sniper is shooting at someone wearing body armour (which is most soldiers today), at a light armoured vehicle or at a target through bulletproof glass, the sniper will require better penetration than is given by the standard low-drag bullet designed for a combination of accuracy and to kill a naked human efficiently. The fitting of an armour-piercing core will also affect the trajectory of a bullet because its centre of mass and average density are changed.

The first bullets were pebbles, or similar, fired from a sling. By experimentation it was discovered that denser materials, like lead, travelled further and lost their speed more slowly, so they hit harder too. Certainly by the 4th century BC the Greeks were casting lead slingshot for use against other people. We have seen in the earlier part of this book how from this point arrows fired from bows took over for some hundreds of years because of technical limitations; but eventually someone came up with the idea of blowing the bullet down a tube with some gunpowder, resulting in far more power behind the projectile. Fired from a musket, a lead ball of 12.7–25.4mm (0.5–1in) could be relied upon to kill or seriously injure a man wherever he was hit, as opposed to one from a slingshot which might only succeed in making him annoyed. Factor in shorter training times, and for military use the firearm was the way forward.

The first musket bullets were spherical, as they were easy to make from lead and could be loaded quickly. Yet the standard musket round would just hit a man at 50–70m (55–77yd) on a good day. Why was the musket so inaccurate? To load into a dirty barrel, and make a gas-tight seal between the barrel and the bullet, the bullet had to be a loose fit to the barrel and wrapped in a paper patch – both had a negative effect on accuracy. Another reason for the paper patch was to hold the bullet firmly against the powder. If a bullet slid away from the powder before firing the detonation was called a 'short start', and often resulted in the barrel bursting. A bad thing on the battlefield or anywhere else for that matter.

Of course the military mind was keen to increase the range and accuracy of the musket and by the 1800s rifling was introduced, although fouling of the bore grooves by the dirty powder prevented widespread adoption. In the late 1700s, cleaner smokeless powder allowed the rifled musket to be used effectively on the battlefield instead of just for sport where nothing was shooting back. The early rifle, however, was still a muzzle-loaded weapon firing a spherical lead ball and accuracy was not perfect, though significantly better than that of the musket.

The next step on the march towards accuracy was the invention, in 1823 by Captain John Norton of the British Army, of the conical bullet. This was still loaded from the muzzle, but when it was fired the gas pressure on the base of the cone caused it to expand and grip the rifling thus avoiding the need for a paper patch. The army rejected this idea, allegedly, because spherical bullets had been used for the last 300 years. In 1836 the famous English gunsmith William Greener invented the Greener bullet. The advance on Norton's design was that the hollow base of the bullet had a tiny wooden plug in it which made it expand more reliably on firing and engage better with the rifling. In tests the Greener bullet proved itself very effective but this too was rejected by the British military as being too complicated to produce in bulk. The next advance was the Minié Ball invented by Claude-Étienne Minié, a captain in the French Army, and introduced by the French in 1847. It was very similar to the Greener bullet, being conical in shape and having a hollow cavity in the rear, but instead of a wooden plug to help the bullet expand into the rifling it was fitted with an iron cap in the rear cavity. When the round

An early sniper – an American marksman hidden in a tree shoots a British soldier during the American Revolutionary War. *(Private Collection / Peter Newark Military Pictures / The Bridgeman Art Library)*

Early Jaeger flintlock rifles c. 1750. *(Courtesy of Martin Pegler)*

Concealed in a tree, American sharpshooter Tim Murphy shoots dead British general Simon Fraser during the Saratoga campaign, in which the British defended Canada from the invading Americans. The American commander Benedict Arnold specifically ordered the targeting of British officers, making this a true sniper operation. *(Painting by Hugh Charles McBarron, US Army Centre of Military History)*

Boer snipers on Spion Kop. The Boers were experienced riflemen, armed with Mauser bolt-action rifles. Their general, Joubert, had bought 30,000 of these excellent weapons together with a quantity of other arms in preparation for this conflict. *(National Army Museum)*

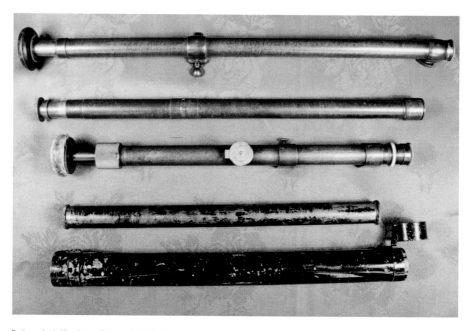

Early optical rifle sights. Telescopic sights have come a long way in the 80-plus years since these were used by snipers on the battlefield. *(Courtesy of Martin Pegler)*

An early scope mounted with the double-claw fixing on a Mauser Gew 98 rifle. Note how heavy the mounting is made to avoid movement of the sight. *(Courtesy of Martin Pegler)*

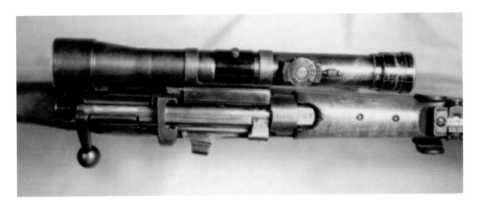

The Mk III Short Magazine Lee-Enfield (SMLE), the most widely used British sniping rifle of World War II. The PPCo scope is offset to allow both the working of the bolt action and for the magazine to be loaded by a stripper clip from above. *(Courtesy of Martin Pegler)*

LEFT: British sniper at his post in the trenches during World War I. The trenches were dug so deep in part to avoid showing soldiers' heads to snipers as they moved around. *(Superstock)*

BELOW: An official Italian photograph showing an Italian sharpshooter sniping at the enemy from behind the double protection of sandbags and a sheet iron counter-sniper plate. Dated 13 July 1916. *(Corbis)*

British snipers from the King's Own Regiment on Salonika in 1916. Note the Galilean (Open Telescope) sights on the rifles resting on the wall and the PPCo offset scope that the nearest soldier has attached to his SMLE. *(Imperial War Museum CO 36)*

A British soldier prepares to take aim as his colleague attempts to draw fire – using a helmet – from a Bulgarian position at Salonika in World War I. *(Getty Images)*

HANDFEUERWAFFEN. I.

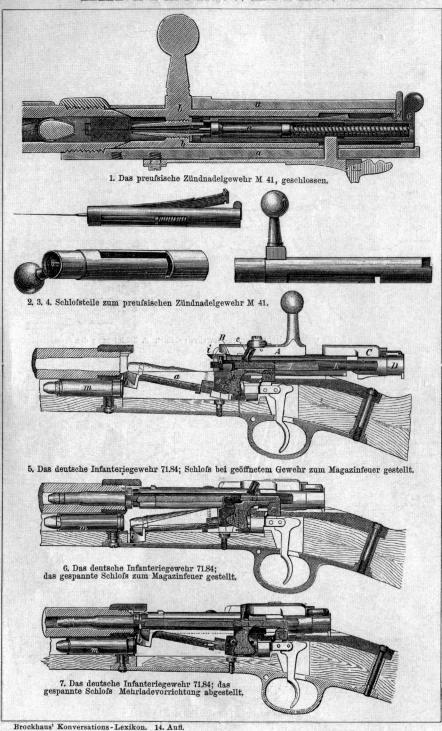

1. Das preußische Zündnadelgewehr M 41, geschlossen.

2. 3. 4. Schloßteile zum preußischen Zündnadelgewehr M 41.

5. Das deutsche Infanteriegewehr 71.84; Schloß bei geöffnetem Gewehr zum Magazinfeuer gestellt.

6. Das deutsche Infanteriegewehr 71.84; das gespannte Schloß zum Magazinfeuer gestellt.

7. Das deutsche Infanteriegewehr 71.84; das gespannte Schloß Mehrladevorrichtung abgestellt.

Brockhaus' Konversations-Lexikon. 14. Aufl.

A German illustration showing a cross section of bolts, magazines and working parts from early bolt-action rifles. *(Corbis)*

An Abyssinian sniper stands in a tree waiting to attack the Italians, c. 1934. *(Corbis)*

These two female Russian snipers (Bykova left and Skrypnikova right), shown in camouflage battle garb and apparently returning from a combat mission, were photographed as part of a propaganda exercise. *(Corbis)*

This Russian solider is armed with the Tokarev SVT-40 sniper rifle, and is likely to have been a sniper on the Eastern Front during World War II. Presumably he has the bandoliers and bayonet for dramatic effect. *(Cody Images)*

Soviet PU scope, mounted here on an SVT-40 semi-automatic sniping rifle. Before long the Germans developed a similar scope, as it was so robust and simple to use. *(Courtesy of Martin Pegler)*

was fired the iron cap very efficiently forced itself into the rear of the bullet, thus expanding the bullet evenly and causing it to engage snugly with the rifling. Finally giving in to the tide of technology, the British adopted the Minié Ball for their Enfield rifles.

The Minié Ball first saw wholesale and successful use in the American Civil War. The war was fought to a great extent with rifled muskets, but it also pretty much marked their end as frontline weapons. They were replaced by the breech-loading rifle in which the whole metal cartridge – consisting of propellant, ignition system and bullet – was introduced to the user-end of the barrel, as this enabled a far faster rate of fire. And rate of fire in many military instances is more important than accuracy, because if one side can fire three times as fast as the other then it is almost as good as having three times as many men.

As soft lead bullets were driven harder and harder down the barrels of early rifles, it was found that when they reached a certain speed they started to wear away significantly against the rifling. At yet higher speeds they began to break up in the barrel. The first answer to this problem was a bullet made from a harder lead alloy. When velocities increased further, however, bullets began to be designed with a brass jacket to give them outer strength, with a heavy lead-alloy core for maximum weight.

From 1862, the British engineer W. E. Metford carried out an extensive series of experiments on bullets and rifling. These led to his invention of a system of light rifling with increasing rates of turn combined with a jacketed bullet. This combination was finally adopted for the British Army in December 1888 as the Lee-Metford .303in rifle, Mark I, predecessor of the legendary Lee-Enfield .303in bolt-action rifle that saw service in British and Empire forces from World War I onwards.

The bullet was now reaching a state of true ballistic efficiency, and scientific understanding of bullet design was also improving, particularly in terms of bullet shape. To minimize the amount a bullet is slowed down by its passage through the air, it needs to be streamlined like a racing car or the hull of a boat. This means a pointed front end but also a tapering rear end to allow the air to curve around it and fill the vacuum behind the bullet smoothly and easily. This tapering of the rear of a bullet is called, not surprisingly, a 'boat tail', and is the design basis of all modern low-drag bullets.

To maintain its speed against the resistance of the air, a bullet needs to be as heavy as possible so that it carries the maximum momentum or kinetic energy. This can be achieved either by making it of a heavier material or by making it longer for the same diameter. To achieve maximum density bullets are already made mostly from an alloy of lead. Tungsten bullets are used in hunting and target shooting, where there is a desire to protect the environment from lead pollution, and depleted uranium bullets are sometimes used for special purposes such as high penetration.

Weight and shape only go part way to giving the bullet stability in flight. The gyroscopic action provided by rifling is also critical. The axis of a spinning gyroscope always tries to stay pointing the same way. Therefore if a bullet is made to spin in its longitudinal axis it no longer tumbles end over end; rather, it keeps the pointed end to the front and cuts through the air like a knife, taking full advantage of its streamlined shape.

It will be obvious by now that bullet design is a scientifically complex business. The modern bullet must be strong enough to withstand the material trauma of being accelerated down a rifled barrel by hot gas to a speed of about 3,219km/h (2,000mph) in the space of a few inches, all while being gripped and spun by the rifling. It must also achieve a certain amount of penetration of a hardened target, yet the jacket must also be soft enough to break up on contact with a human body to do maximum damage.

The effect of a bullet hitting a human body is similar in some ways to an aircraft hitting the sea at speed. The inertial resistance of the target produces an impact similar to hitting a solid wall, but the composition of a human body is not exactly like water and varies according to the flesh, bone and clothing etc present at the point of impact. This means that there can be a wide range of effects produced when an identical bullet hits different people at different ranges.

The bullet must be designed to achieve optimum destruction in all cases. If a bullet does not break up when it hits a person, then it can sometimes drive a tiny hole right through without doing sufficient damage. If the bullet breaks up into fragments, or expands its surface area, it will rip the body inside and consequently cause greater blood volume loss, the primary mechanism of bullet stopping power. When a bullet hits a soft target it might usefully be imagined as a wedge that pushes the matter at the impact area laterally, relative to its path, with great violence. This creates a lateral shock wave which makes the target effectively explode outwards from the point of impact, like hammering a wedge into the end of a dry log at high speed. A larger-calibre bullet, travelling at the same speed, will move more target material further sideways than a smaller-calibre bullet. This is one of the reasons why sometimes the 5.56mm round does relatively little damage beyond a piercing hole, while the 7.62mm round often produces an exit wound the size of a fist and the 12.7mm round tends to immolate the target as if it contained explosive.

A high-velocity bullet also projects a cone-shaped shockwave in front of it as it passes through a human body, and this shockwave does tend to destroy the internal organs whatever the deformation of the bullet; water-filled organs such as the bladder are particularly vulnerable to the shockwave. Indeed the shockwave will often still do serious damage when the bullet is stopped by flexible body armour as the shock wave will pass through the armour into the body.

Very often nowadays the sniper's quarry is going to be wearing rigid body armour or sitting behind armoured glass or steel plate. The normal sniper's bullet will break up and not penetrate this type of protection, so something with more penetration is required. An armour-piercing bullet designed for a rifle normally either has a hard metal tip, perhaps made of tungsten carbide like a chisel, or it has a thin bar of this metal like a nail running along its centre line. Then, when the bullet hits the hard target, the outer case is stopped or slowed but the coaxial bar drives a small hole through the hardened target. While this is not perfect for inflicting maximum blood loss and other damage, it is often sufficient to kill the target.

If an armour-piercing round is fired at an armoured vehicle, the part that penetrates the armour might be expected to bounce around inside the crew compartment and damage the occupants. Unfortunately for snipers, the designers of armoured fighting vehicles (AFVs) often line vehicle interiors with a material that catches such items. The armour-piercing bullet core may hit one occupant, but it will not bounce around to optimum effect. To get around this problem, there are armour-piercing rifle bullets designed to explode within the confined space of an AFV. As it takes a fairly large bullet to carry a worthwhile explosive charge, the calibre is usually 12.7mm

(.50in) or more and fired from a rifle designated 'anti-materiel'. Certainly a .50-cal bullet packed with high explosive detonating inside a confined space will kill the occupants owing to the pressure wave of the detonation being concentrated inside the vehicle.

In addition to armour-piercing ammunition, the sniper can also make use of tracer bullets, first introduced by the British in 1915. The purpose of a tracer round is either to point something out, like a target, to someone else or to observe the fall of the shooter's own fire. The tracer bullet has chemicals inside towards the slightly extended rear, usually a mixture of magnesium metal, a perchlorate and strontium salts, which are ignited by firing the round. Once the chemicals are ignited, they give off light, generally red, when the bullet has travelled a couple of hundred metres. (The delay in ignition is so that the shooter does not give his position away so much.) The tracer is very effective for target designation, but not so good for a sniper checking the fall of his shot, as the tracer is longer and less boat-tailed than a standard bullet, plus it reduces in weight as it travels, so its flight is not precisely the same as a conventional bullet from the same weapon. For this reason the tracer is not very useful for the sniper but it is invaluable for observing the fall of shot from a machine gun or marking targets for a fire team. Working along similar lines to tracers, incendiary bullets are made with an explosive or flammable mixture in the tip, which is designed to ignite on contact with a suitable target. The idea is that they are used to ignite fuel or munitions in the target area.

Note that under international law the use of certain types of ammunition against humans is banned, although as far the author is aware, little notice is taken of these rules by the military of any nation except *after* a conflict, where infraction is used to convict the losers of war crimes. The St Petersburg Declaration of 1868 prohibited the use of explosive projectiles weighing less than 400g (14.1oz). This rule appears to be totally ignored given that the US Army has just issued the XM25 rifle, which fires a 25mm (1in) explosive bullet containing electronics that allow it to be made to explode inside a room or over a trench. Protocol III of the 1983 Convention on Certain Conventional Weapons, an annexe to the Geneva Convention, prohibits the use of incendiary munitions against civilians but not against soldiers. The use of certain kinds of ammunition by soldiers against the uniformed military personnel of opposing forces is prohibited by the Hague Convention. These munitions include projectiles that explode within an individual, and poisoned and expanding bullets. Nothing in these treaties prohibits tracers or the use of prohibited bullets on military equipment, by which is presumably meant vehicles not men.

The Cartridge & Propellant

For a sniper, premium quality ammunition is essential, and cartridges are selected on the basis of exacting requirements. In essence, a cartridge is a tube (usually metal) full of propellant with a bullet at one end and an ignition system at the other. The use of a metal cartridge is closely associated with the principle of loading from the breech of the weapon (as opposed to the muzzle), and the far higher rate of fire which this combination allows.

By the 1860s, the muzzle-loading rifled musket, firing the Minié bullet, had come about as far as it could in terms of accuracy and rate of fire. It was shooting a pointed bullet with clean(ish) smokeless powder down a rifled barrel to a range of several hundred metres, and at a rate of about three or four shots a minute. Its firepower was limited only by the speed with which the rifleman could go through the procedure of pouring powder down the barrel followed by a bullet and so on while being shot at.

Spreading rapidly across the world by this time, because it had a rate of fire so much higher and could be reloaded lying down, was the replacement system: a breech-loading rifle utilizing a brass cartridge that combined the bullet, propellant and primer in one easy-to-use package. It was loaded into the breech of the weapon and produced an accuracy and rate of fire so superior to muzzle-loaders that it rendered them obsolete almost immediately it appeared on the battlefield. Breech-loading cannon had been in use since the 1400s, but they didn't employ a cartridge, as such; they had a number of breech-parts shaped like drinking mugs which were wedged into position at the breech end ready for firing in the normal way with a lighted taper applied to a touchhole. Leonardo da Vinci, in his work *Codex Atlanticus*, written between 1478 and 1519, demonstrated one of the main types of breech-loading systems that persisted until the invention of the enclosed cartridge in the 19th century. The breech end of the harquebus, or heavy musket, was screwed off and an oversize ball was placed in the barrel. (This procedure allowed for a snug fit in a smooth bore and a good grip of the grooves later when rifling was invented.) The powder was added and the breech screwed back on ready to be primed and fired with a match or taper. This procedure was cumbersome and slow at best, and by the end of the 1700s many other types of breech-loading system had been tried – mostly based on the idea of a threaded hatch in the top of the barrel rear through which a ball could be inserted followed by the powder propellant. The advantage they all offered was the accuracy produced by a snug fit of the bullet in the barrel, but they all remained tediously slow to operate and were regarded with contempt by the military authorities on account of their battlefield fragility. (For the civilian reader, soldiers are well known for their predisposition to wreck or misuse any piece of equipment no matter how strong, so equipment for soldiers is routinely over-engineered and then tested with a small unit of men to discover its weaknesses before it goes into mass production and subsequent general issue.)

Though ammunition for muskets had been supplied previously in card cartridges, these were for convenience only, as they had to be bitten open and the powder poured down the barrel followed by the ball when loading. It was not until 1808 in Paris that French gunsmiths Jean Samuel Pauly and François Prélat developed the first self-contained cartridge intended to be loaded at the breech. This was made of thick paper with a copper base and held a round bullet. This cartridge was designed to be loaded through the breech of theweapon and fired by a needle (the firing pin principle) striking into the primer contained within the copper base. Both this needle-firing concept and the idea of the self-contained cartridge remain the basis of all firearms to this day.

The lack of an effective gas seal where the breech fitted to the barrel prevented many early breech-loading weapons from being widely accepted. This problem was solved, however, by the same Parisian gunsmith Jean Pauly, mentioned above, and one of his staff, Johann von Dreyse. Dreyse's invention was called the 'needle gun' and utilized a bolt-action breech-locking system that sealed the chamber effectively and fired a bullet from an efficient self-contained cartridge. This system was adopted by the Prussian Army in 1841, making the Prussians the first army in the world to use a breech loading, self- contained cartridge system. Once Dreyse's cartridge and bolt-action system came onto the scene other weapons became obsolete overnight owing to the massive increase in rate of fire which the cartridge system allowed. Even without magazine feed, the single-shot bolt action was many times faster than any muzzle-loader. Another very important advantage of the breech-loading rifle over the muzzle-loader, to a soldier at least, is the fact that it can be easily loaded while in the lying position and thus while presenting a small target to the enemy. Of course, every aspiring inventor of weapons was looking to improve the system yet further. By about 1870 metal cartridges had proven their superiority over paper due to their rigidity when handled by bolt or 'working parts', and many types of brass-jacketed cartridges were in use. A few cartridges had the primer in the rim of the cartridge rear and were called 'rimfire'

A modern cartridge consists of the following:

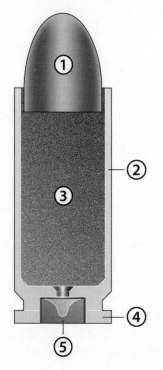

1. the bullet, which serves as the projectile;
2. the case, which holds all parts together;
3. the propellant, for example gunpowder or cordite;
4. part of the casing used for loading;
5. the primer, which ignites the propellant.

cartridges, while most had the primer in the centre and were called 'centrefire'. All military weapons now use centrefire cartridges. Primers in this position are less liable to accidental detonation when roughly handled and a central primer allows the rim of the cartridge to be stronger and avoid deformation from the combustion of a high-pressure load.

All cartridges are essentially bullet-delivery systems, and the speed at which the bullet leaves the muzzle is called the muzzle velocity. The higher the muzzle velocity the flatter the bullet will travel and the more damage it will do to the target. Travelling faster also reduces the deflection in flight caused by a crosswind. To achieve the high muzzle velocity required of a sniper rifle, the propellant must burn at a controlled rate – not so quickly that the weapon explodes in a detonation or that the bullet is destroyed as it enters the rifling, and quickly enough that, by the time the bullet reaches the muzzle, it is moving at the desired velocity. To do this the cartridge must be large enough to contain and burn the right amount of the right type of propellant.

A modern sniping rifle has a muzzle velocity of about 750–900m/sec, which equates to roughly 2,500–3,000ft/sec. The speed of sound is 315m/sec (1,033ft/sec) at sea level though this varies quite a lot with air temperature. The speed of the bullet falls off with distance as the air slows it down. The maximum velocity to which a projectile can be accelerated in a barrel is limited by the expansion rate of the gas from the burning propellant. The practical limit for a normal bullet is 1,200m/sec (nearly 4,000ft/sec) achieved in small-calibre guns that can fire light bullets, about the same as achieved by some artillery cannon firing conventional heavy-calibre high-explosive shells. The highest velocities achieved by any firearm are by tanks firing armour-piercing discarding-sabot (APDS) rounds that reach 1,800m/sec (6,000ft/sec) which is approaching the theoretical limit for powder propellants. To drive a solid projectile much faster than this would require different technology such as an electric railgun.

The terrific pressure – in excess of 3,515kg/cm^2 (50,000lb/in^2) – which tries to burst the cartridge every time a round is fired in arifle can potentially cause the cartridge body to swell and stick in the breech. One idea to prevent this happening is to introduce a taper to the length of the cartridge body so that as it moves rearwards from the breech during the extraction process it necessarily comes loose. This principle is employed in many Russian military cartridges such as the AK-47 assault rifle 7.62 x 39mm and the PKM machine gun 7.62 x 54mm. There is a very slight, invisible taper on NATO rounds, but all Soviet ammunition has a very obvious taper to assist the release from the breech.

Another issue affecting cartridge design is 'bolt thrust'. This is the push on the gun's bolt, or breech-block, towards the rear of the weapon caused by the cartridge firing. The strength of the bolt thrust depends on two factors: the pressure generated within the chamber and the diameter of the base of the cartridge case. The wider the cartridge case the greater the surface area the pressure inside the cartridge has to push against, so the greater the bolt thrust for any given chamber pressure. So, while the same chamber pressure might be developed within a long thin cartridge case as within a short fat one, the latter will develop double the bolt thrust, because it has double the internal base area. And the greater the bolt thrust, the stronger the locking mechanism restraining the working parts has to be to withstand it. As far as the user is concerned this tends to make the rifle heavier.

Bolt thrust is part of what constitutes recoil, the 'kick' delivered to the shooter's shoulder as a round is fired. Recoil stems from the basic physics of Newton's Third Law Of Motion: 'Every action has an equal and opposite reaction.' What this means is that the weight of a bullet multiplied by the speed with which it leaves the end of the barrel not only equals the thump it is going to give to the target (less its loss of energy through air resistance), but it also equals the push backwards it is going to give to the rifle firing it.

The more powerful the rifle, the harder the recoil, a fact that can be very important to a sniper. If a sniper wants to hit a target as far away as possible, then he wants the heaviest bullet travelling as fast as possible from the muzzle of his rifle which, all other things being equal, means the biggest kick the sniper can handle. Generally speaking, it is not practical to fire anything much bigger than a .50-cal sniping rifle or the shooter will get 'gun shy' from having a bruised or broken shoulder. With 5.56mm rifles or the AK-47 the kick is not sufficient to be concerned about, but with a .50in Barrett or similar the shooter is liable to suffer severe bruising. And there are even heavier weapons planned for use in the anti-materiel role.

Fortunately there are a couple of things which can be done about this recoil problem. The first is to use a heavier rifle because, while a heavier weapon does not actually *absorb* the energy of recoil, it does slow down the resulting movement. An example will make this plain. A rifle weighing 4.5kg (10lb) will be driven back by the recoil force at half the speed of a rifle weighing 2.27kg (5lb) after firing the same round. The second way of reducing the effect of recoil is something called a muzzle brake. Following the bullet out of the muzzle of the rifle is a stream of hot gas moving very quickly in the same direction. The muzzle break is an arrangement on the end of the muzzle which catches and deflects the gas coming out of the muzzle to reduce the recoil. It is a steel moulding mounted on the muzzle but set forward of it with a hole for the bullet to travel through. Around that hole is a surface for the gas, which is expanding laterally as it travels, to hit and push upon so that it is deflected to the side or rear of the muzzle. The effect of the gas being deflected is a force tending to push the rifle forward against the recoil and this force reduces the kick by up to 30 per cent. Something to be aware of when sniping with a rifle fitted with a muzzle brake is that the deflected gas can kick up dust around the muzzle and give away the shooter's position. Heavy-calibre sniper rifles are often fitted with a combination of suppressor and muzzle brake to reduce this problem.

As a final measure to reduce recoil, the designers can utilize various systems of buffers and recoil springs to absorb the recoil more effectively, transferring less of it into the shooter's shoulder, or at least ensuring that the same force is spread out over more time in a controlled manner.

The Barrel & Breech

Today, all quality rifle barrels are drilled from forged bars of steel. The steel type varies according to the requirements of the weapon and the need for price control, but it has to be of high quality, with no impurities to cause localized weakness. The main type of steel used in making the barrels of hunting rifles and military firearms is high alloy chrome molybdenum, which is the same type of steel as is used in engine con rods and truck axles. Target shooting barrels are mostly made from stainless steel, specifically a martensitic type which is more accurately described as 'free machining' or 'rust resistant' because it has a chrome content of around 10 per cent. Unlike true stainless steels, martensitic steels can be hardened by heat treatment like regular carbon steels and this makes the production of a strong, wear-resistant barrel possible. The best sniping rifles in the world are made with 416 rust-resistant steel. Another benefit of this steel is that if the user follows the instructions in the owner's manual to correctly 'shoot in' the rifle, then the inside of the barrel will acquire a burnished polish which pretty much eliminates fouling from the propellant, so it will be easy to clean.

We have already discussed the fact that a barrel must be rifled – fitted with helical grooves along its bore – to make the bullet spin and fly true rather than tumbling. But the number and gradient of these grooves is very important for their effect on the range and accuracy of the bullet. As a rule, rifles under 7.62mm/.30in calibre have four grooves and those above have six. The principal reason is that the bigger, and therefore heavier, the bullet, the harder it is to get it spinning rapidly against its inertia, so there need to be more grooves for a better grip. The idea is that the bullet must spin fast enough to stop it tumbling but no faster as more spin equals more resistance from the grooves of the rifling, which slow the bullet down. Also, if a bullet is made to spin too quickly it can fly apart from the centrifugal force – there are some serious forces at play here, as a rifle bullet will be spinning at 100,000–300,000rpm when it leaves the barrel.

A fact of physics is that longer bullets need to spin faster than shorter bullets to remain stable. Fortunately we do not need to work out the maths here, but to give a general idea of what is involved, a bullet leaving a barrel at 914m/sec (3,000ft/sec) where the barrel twists once per foot would be rotating at 180,000rpm. The slowest rifling 'twist' is found on muzzle-loading firearms. A muzzle-loading rifle might have one turn in 122–152cm (48–60in), whereas the M16 5.56mm rifle has one twist in 17.8cm (7in). As a general rule, smaller calibres have higher twist rates. A very high-velocity sniper rifle, firing a hard-to-stabilize long bullet (for a low drag coefficient) is going to have a great deal of resistance from the inertia of the bullet to overcome – its resistance to spinning. Remember the bullet is going from stationary to a forward speed of about 800m/sec (2,625ft/sec) and a spin of 300,000rpm in a distance of a metre and in a time-span of less than 1/800th of a second. To help get around this inertia problem, the best and latest rifles have shallow rifling at the breech end, so the bullet starts spinning relatively slowly, then the pitch of the rifling increases along the length of the barrel until it reaches the correct pitch, and therefore spin, for the requirements of bullet stability.

There are three main ways of putting rifling in a barrel. The oldest and slowest, but the best, way is to cut the rifling by sliding a cutting tool backwards and forwards along the barrel. This is called, reasonably enough, 'cut rifling'. The next method to be invented, and almost as good but quicker, is 'button rifling' where a 'button' tool is actually pulled the length of the barrel once and during this trip it melts the steel inside and forms the grooves.

The third method is 'hammer rifling', which is very fast but not quite so accurate as the previous methods. A machine forges the barrel from hot metal around a mandrel or former with the rifling pattern on it. This is the type of rifling production method which the Germans invented during the war for the rapid production of the MG 42 machine-gun barrel.

The chamber is the larger diameter part of the barrel interior, at the opposite end of the barrel to the muzzle, which houses and holds together the cartridge case for firing. There is more pressure pushing outwards around the chamber of the barrel, because this is where the initial burning of the propellant occurs, and therefore the highest pressures. Note that some target shooting barrels continue at the same thickness all along their length to improve stiffness, but most military weapons taper towards the muzzle to reduce weight.

The length of the barrel bears a complex relationship to the calibre of the round being fired, and a simple one to the type of propellant. The barrel has to be long enough for the rifling to get the bullet spinning at the speed required for stability. With a sniper rifle firing a heavy bullet that may mean a long barrel. A barrel also has to be long enough to extract all the energy from the propellant in the cartridge; assuming the cartridge is correctly loaded. This generally means a long barrel if a high muzzle velocity is required. A cartridge with more propellant than is correct for the length of barrel sends a lot of flame out of the end of the barrel and gives away the position of the shooter. If a barrel is too long relative to the propellant charge, however, it will cause the gas pressure behind the bullet to drop towards the muzzle and this will allow friction to reduce the muzzle velocity. Though it might produce a higher muzzle velocity or allow more rifling for a faster bullet rotation, the one thing a long barrel alone *does not do* is make the rifle more accurate.

At the business end of the rifle, the muzzle of a rifle has to be carefully machined so as to allow the bullet to leave cleanly. As this is the bullet's last point of contact with the barrel, a slight inaccuracy here would throw the bullet off course or cause it to topple in flight. A sniper rifle generally has a barrel length of 61cm (24in), or longer, to allow the cartridge propellant to burn fully. Besides increasing the bullet's muzzle velocity, this also reduces the revealing muzzle flash mentioned above. Most standard military rifles are also fitted with a slit-type flash suppressor mounted on the muzzle to reduce the muzzle flash at night, in an attempt to reduce the shooter's visibility to the enemy, a measure that is only slightly effective. Far better is a suppressor that is often known incorrectly as a 'silencer'. It is a tubular container that screws to the end of the barrel and allows the gas leaving the muzzle to expand within it, thus capturing much of the noise, and flash, created when the rifle is fired.

It is not possible to make a normal sniping rifle silent, as the supersonic crack of the bullet passing through the air tells everyone in the target area that someone is shooting in their direction. The point of a suppressor, however, is that these people around the target cannot hear where the shooting is coming from, so they cannot so easily work out where the sniper is situated. Occasionally, on a clandestine mission, a sniper will be issued with a large-calibre sub-sonic sniping weapon, perhaps a modified Heckler and Koch MP5 DS designed to fire a 9mm bullet at just under the speed of sound. This round can be silenced very well indeed and is accurate up to around 274–366m (300–400yd). The relatively heavy round travelling slower than the speed of sound can still pack something of a punch, but makes no sonic boom to give the shooter away.

As already noted, the barrel of a rifle flexes every time a round is fired. Only a tiny amount of course, but we have seen how a tiny amount over the length of the barrel makes a huge difference at long range. To make things a little more interesting, the barrel does not whip just up and down. Owing to the reciprocal effect of the rifling on the inertia of the bullet, the muzzle describes an ovoid shape so the inaccuracy is in all directions vertically and horizontally but not equally.

Clearly this is a problem when designing a rifle, and it cannot be eliminated completely, but it is possible to reduce barrel whip by having a thicker barrel, which is why most target rifles have thick barrels. The trouble is that this makes the barrel a whole lot heavier for a sniper to carry, so very often a compromise is reached by making the barrel fluted. This is where there are 'fins' running the length of the barrel. Like an 'I' beam girder, the fluted configuration is the strongest design for minimum weight. In any event, to achieve maximum accuracy the bullet must leave the muzzle at the same stage in the barrel's wave motion on every shot. Many factors affect the pattern described by the muzzle during this wave, the most obvious of which is the pressure exerted on the barrel by the inertia of the stationary receiver and the flash suppressor attached to the fore-end. Support under the barrel, for instance, alters the barrel whip and changes where it is pointing when the bullet leaves the muzzle, in the same way that holding down a string on a guitar shortens the length of string which is vibrating and thereby alter its frequency of vibration. Placing a fore-end support under a barrel will shorten the length of barrel that vibrates, and thereby alters the direction the muzzle is pointing when the bullet leaves it. The reader will observe that many competition and sniper rifles now have a totally 'free-floating' barrel which does not touch the fore-stock or any support.

Of course there are other ways to get around the barrel tuning problem. As an alternative to having a free floating barrel, connected only to the receiver and tuned as such, there may be a fixed support under the barrel which is always the same – usually the bipod. By having the bipod supporting the rifle every time, the barrel whip is the same every time and the sights are set up to suit this. The well-trained sniper, of course, is very aware that his fall of shot will change according to where the barrel is supported.

Loading Mechanisms

Loading mechanisms are another area of rifle design extremely important to the sniper, and they constitute a subject that generates its own field of controversy amongst the sniping community. As we have seen, once the metal cartridge and breech-loading system had been invented the process of loading speeded up considerably. The first type of breech-loading action to be employed with the associated metal cartridge was the bolt action, followed quickly by the underlever action. Almost immediately, these two 'manual' actions, so called because a lever has to be operated to extract the empty cartridge case and then, in repeated use, load the next cartridge, were fitted with magazines to increase their rate of fire. In all but the very first examples, working the action also cocked the firing mechanism, pulling back the hammer or compressing the spring which drove the firing pin.

The first underlever action rifle, the Spencer Repeating Rifle designed in 1860, was issued to some Union troops during the American Civil War. It had a seven-round magazine of .52in rounds located in the butt and a fresh cartridge was introduced to the breech with each operation of the lever, but the firing hammer had to be cocked separately. This allowed a rate of fire of 14–20 rounds per minute, five times that of a musket. The British Army adopted the Martini Henry underlever a little later and this was used at the battle of Rorke's Drift in South Africa, 1879, where just over 100 British infantry held off 4,000 Zulu warriors.

It didn't take long for some inventors to realize that a weapon which used the power generated by the firing of the propellant to eject the empty cartridge case and reload the rifle after each shot would increase the rate of fire dramatically and produce a type of internally powered machine gun. The hand-cranked Gatling Gun had been around since 1861, so the advantages of a high rate of fire were already well known. In 1887 General Manuel Mondragon invented the Mexican Mondragon Rifle, which was a 7mm-calibre, gas-operated Mauser supplied initially with an 8- or 20-round magazine. This weapon had great hitting power, but the recoil was severe and it suffered terribly, as do many automatic weapons, from stoppages due to dirt in the working parts. The weapon had a fire selector switch allowing both semi-automatic and fully automatic fire and produced an amazing cyclic rate of fire of between 750 and 1,400rpm. Later developments were fitted with 100-round box magazines and, with the (similar) American Browning Automatic Rifle (BAR), were the first automatic rifles ever to see service: during World War I in 1914–18. It had taken over 25 years for the US military establishment to accept the idea of an automatic rifle. To be fair, it took the British top brass until 1957 to issue a semi-automatic rifle. In defence of both sets of leaders, there is an argument that a soldier equipped with an automatic weapon tends to waste ammunition in the excitement of combat and can end up running out of this vital commodity.

There is an important distinction to be made between semi- and full-automatic weapons. A semi-automatic weapon fires and reloads once each time the trigger is depressed, while a fully automatic weapon fires and reloads and fires again continuously for as long as the trigger is held down, or until there is no more ammunition in the magazine or on the belt. (Note that so-called automatic pistols are generally not automatic but semi-automatic. The fully automatic pistol is normally fitted with a stock and called a submachine gun in English or, more accurately in German, a machine-pistol.)

The various mechanisms of self-loading and automatic weapons, or at least those relevant to a sniper, are as follows:

Recoil

The energy developed by the recoil of the barrel is used in some weapons to operate the working parts. In most types of hand weapon, the whole weapon recoils when a shot is fired because the barrel is mounted solidly on the receiver. In a recoil-operated weapon an arrangement is made for just the barrel to recoil against the receiver of the weapon and the relative movement is used to operate the working parts. This system is not used very often on modern rifles, but was employed on the Remington Model 8 semi-automatic rifle (1906) and many shotguns. The Browning .30- and .50-cal machine guns are probably the best-known modern Western examples of recoil action, but there are many types of recoil-operated machine guns in the Russian and Chinese arsenals. This system, while producing a relatively slow cyclic rate of fire, is very reliable.

Gas

Gas operation is the most common system by which rifles are reloaded automatically. A little of the gas driving the bullet down the barrel is tapped off through a hole and used to operate the working parts. This can be done in one of three ways. Most simply (called 'direct impingement'), the hole in the barrel can funnel the gas back down a tube parallel to the barrel directly to the working parts and drive them back. The disadvantage is that the gas makes the working parts dirty very quickly, but the advantage is that it reduces recoil dramatically and, by reducing the working parts, limits the possibilities of mechanical failure. Weapons which use this system include the Swedish Ljungman Ag m/42, the M16 and the French MAS-49. The second variation of gas operation, called long stroke, uses the bled-off gas to drive a piston backwards. This piston rests against the working parts and pushes them all the way back as far as they are required to travel. The American Garand rifle and the Belgian FN 7.62mm rifle used the long stroke system. The third variation is called short stroke, and here higher pressure gas is tapped off a little closer to the breech. This gas drives a short piston, which is already resting against the working parts, back a short way with sufficient force to cause the working parts to continue back as far as required by their own momentum. The M1 carbine, the M14 rifle and the M60 GPMG all use the short stroke system. The main advantage of this system is that it reduces the total weight of the working parts and thereby the recoil but it does also tend to produce a higher rate of fire than a similar-calibre weapon operated by recoil.

Blowback & blow forward

Blowback operation utilizes the gas pressure pushing the cartridge case backwards at firing to operate the working parts, via bolt thrust. This system is limited as to the gas pressure it can handle by the strength of the cartridge case, so it tends to be used in low-pressure systems which fire pistol ammunition. These include automatic pistols, submachine guns and semi-automatic shotguns.

Blow forward action is where the friction of the bullet travelling down the barrel pulls the barrel forward on firing and is harnessed to extract the empty cartridge case and reload. This means the barrel is almost the only moving part and therefore offers potential reliability. The first weapon to use this system was the Steyr Mannlicher M1894 pistol of Austro-Hungarian origin and the first rifle was the Swiss

SIG AK53. The only time this system has been used by the modern military was in a 40mm automatic grenade launcher issued to the US Navy during the Vietnam War. There is a modern design and patent for its use in an automatic 12-gauge shotgun called the Jackhammer, but no one seems keen to manufacture it on any scale.

The main problem, or complaint against, automatic weapons has always been their lack of reliability, as a little grit in the working parts brings the loading process to a halt very quickly and can leave a soldier virtually unarmed. Besides this, an automatic weapon is going to fire a lot more rounds than a manually operated equivalent and therefore a lot of carbon from the propellant will build-up around the working parts, leading to possible jams and malfunctions. By World War II the Russians had adopted the principle of designing weapons with loosely fitting parts to allow for a lot of grit and carbon to enter without stopping the action, albeit at the cost of marginally reduced accuracy. The answer to dirt-induced stoppages in all weapons is regular cleaning and this must be enforced through regular inspection.

For the sniper, there are issues with all the above mechanisms. The moving parts of gas- and recoil-operated weapons cannot help but make such rifles slightly less accurate than a bolt-action rifle, and blowback cannot be used on a sniping rifle at all, on account of the risk of a high-power cartridge bursting. A sniper must make the choice between being able to take down several targets in quick succession, or make rapidly repeated shots at the one target, with a semi-automatic rifle or opt for the best accuracy available with a bolt-action weapon. Where a sniper is operating in the true sniper role and taking long shots over open ground, a bolt-action weapon is generally considered to be the best option. Where a sniper is operating more as a squad marksman or 'designated sniper' within an infantry section, his job is really to increase the reach of the section by a few hundred metres rather than take down targets at more than a kilometre. In this position he will have to add his weight to the general fire power of the section as required too, so a semi-automatic is certainly the best choice. This weapon will often fire the same round as the rest of the section – a useful factor when it comes to sharing ammunition – and may indeed be the same weapon as they are using, perhaps with a fitted stock and better telescopic sights to raise the level of accuracy.

Stocks, Grips & Supports: Stability & Comfort

The stock of a rifle is the wood or plastic material surrounding and supporting the barrel and the receiver and running back to the shooter's shoulder. The primary function of the rear portion of the stock, or butt-stock, is to allow the shooter to set his aiming eye at the correct distance behind the scope or sights to use them efficiently and accurately. This distance is called eye-relief. It also gives him a support to rest his cheek on to steady the position of his eye laterally relative to the eyepiece of the telescopic sight. The secondary purpose of the stock is to reduce the effect of recoil on the shooter's shoulder and cheek. The third purpose of the stock is to enable the shooter to grip the weapon firmly and hold it steady while shooting. Finally, the fourth purpose is to stop the shooter burning his hands on a hot rifle barrel. (Note that the frontal part of the stock around the barrel is commonly referred to as the fore-stock.) A sniper does not normally fire rounds quickly enough to get the barrel very hot, but a designated marksman is very likely to do so.

The most common features that differentiate a sniper rifle stock from that of a combat rifle are typically an adjustable cheek piece, which can be adjusted to place the shooter's eye level with the telescopic sights, and an equally adjustable butt length to provide the correct eye-relief. Both the part of the butt and the cheek pad may also be made of a material with recoil-absorbing properties. On bolt-action sniping rifles, some shooters prefer a stock with a thumb hole to aid their grip without raising the weapon higher than necessary above cover. This may be moulded in such a way as to give a purchase similar to a pistol grip.

In addition to the stock itself, there are various accessories and attachments a sniper can use to improve the stability, and therefore accuracy, of the shot. A bipod, or even a monopod, fitted to the fore-end or barrel, makes a much more stable base for the front of the rifle than a shooter could ever create with his hand, so folding versions often come as standard issue with sniper weapons. Some rifles also feature a rear monopod, which forms the third leg of a tripod in conjunction with the front bipod, and thereby gives the shooter an extremely solid firing position. Very often it cannot be employed owing to the shooting position not being suitable, but it should be used whenever possible as it does increase steadiness and accuracy dramatically.

Many infantry rifles have, for generations, been fitted with carrying slings strung between the rear or the butt-stock and the fore-stock. The use of a sling to improve shooting is an old fashioned, but effective trick. Where a bipod is not available a shooter can make his grip on the rifle, particularly at the front end, much more rigid by attaching a sling to the fore- and rear-stock and then wrapping it around his forearm in such a way that it pulls the rifle into a tight lock against his forearm, shoulder and cheek. This used to be called the 'Bisley Sling' owing to its historical use at the Bisley competition shooting venue in the UK.

Iron Sights & Telescopic Sights

The whole idea of shooting is to get the bullet to strike as close as possible to where the impact is required. Shooting at different ranges, particularly when the target is an enemy soldier, makes different calls on the sighting system. When the target is close up, accuracy is not so important as the target seems relatively large, but the target may be preparing to shoot at the shooter, so speed could be vital. When the target is a long way off, as is often the case in asniping situation, the shooter often has more time to take the shot, but the importance of accuracy increases, simply because it is more difficult to hit a target further away. Some types of sights, such as telescopic ones, are accurate but slow to bring to the aim. Iron sights, essentially metal posts and apertures on the top of the barrel and receiver, are much quicker to bring onto the target but less accurate than optically enhanced sights. To take the conflicting constraints of accuracy and aiming speed requirements into consideration, a range of different sight types have been developed. As a general rule, telescopic sights are always used for long ranges, whereas iron sights or some other clever technical models which give a wide field of view are used in short-range combat.

In terms of the actual types of sight a sniper might use, iron sights are by far the simplest. They are referred to as 'iron sights' for the simple reason that they are metal and have no glass optical parts. Open iron sights consist of a post at the muzzle end of the barrel (front sight) and a notch of some kind at the breech end (back sight). Open sights are the fastest to use of all rifle sights and are sometimes referred to as 'battle sights' because, in a close-up gun battle, the shooter is not going to be shooting much above the 100m (109yd) range to which they are accurate. Though most pistols only have open sights, very often a rifle is equipped with open sights in such a way that the rear sight folds up in some way to change into an 'aperture sight' (see below).

With iron open sights the shooter aims by lining up his eye behind the rear sight in such a way that the front sight post sits in the middle of the back sight notch and level with the top of that notch. Then the rifle barrel is moved until this sight pattern has the target just sitting on the post. This combination is called the 'sight picture'. From the shooter's point of view, there should be a space between either side of the front sight and the edges of the notch. These spaces are called light bars and balancing their brightness on either side provides the shooter with feedback on the accurate alignment of the post in the notch. To adjust a set of open sights the rear notch is usually mounted so that it can be moved left or right by a screw thread. Move the sight left, for example, and the bullet strike point will move to the left.

A more advanced form of iron sight is the aperture sight, also known as a peep sight, which is used for ranges of up to several hundred metres, depending on the skill of the shooter. It consists of a post or blade front sight and a small-diameter aperture for the rear sight. This set up is more accurate than the open sight, but slower to bring to the aim.

The more accurate aperture sights are made with tiny holes in a plate of metal and the smaller the hole, within reason, the more accurate they are. In the world wars, soldiers aiming for increased accuracy would sometimes put some thin material over the rear sight aperture and make a pin prick in it to diminish the size of the visible hole. Slightly less accurate, but quicker to bring to the aim, are 'ghost ring' aperture sights, which have a rear sight consisting of a thin ring. In use this guides the rear eye, but blurs into near invisibility (hence the name 'ghost'), facilitating quicker aiming than a normal aperture sight. Aside from the facility to adjust the rear sight with a screwdriver for zeroing, the rear sight of an aperture sight very often has an arrangement by which the shooter can quickly raise or lower it relative to the barrel by sliding it up a 'ladder' or turning a graduated screw thread so as to set the range in hundreds of metres. Aperture sights are used on target rifles and most military rifles such as the M1 Garand, the Lee Enfield and the M16 series of weapons.

The next most accurate sight is the red dot or laser sight, in which a laser beam coaxial with the barrel of a weapon projects a red laser dot onto any object with an effective and accurate range of about 50m (55yd). The idea is that in dim conditions, such as indoors, the shooter can just lay the red dot onto the target and pull thextrigger. Obviously, this is more applicable to pistols and submachine guns than long rifles. Some variants of this sight can also be used when shooting in bright and sunny conditions by looking through a viewing window on the weapon.

Open sights have many advantages: they are cheap to produce, quick to aim, simple to use, soldier proof, lightweight, resistant to cold and damp, and they do not require batteries. Yet they are simply not as precise as other forms of sight. Even a tiny error in the angle of sight alignment results in a bullet path that diverges from the target by a significant margin. And the shorter the distance between the front and rear sights, the more difficult this error is to control. For example, the site radius, or distance between front and rear iron sights, on a military rifle might be 660mm (26in). If the rear sight is only 0.2mm (0.008in) out of line then at 1,000m (1,094yd) that misalignment is magnified 1,500 times and results in the shot being off by 30cm (12in).

Iron sights have other deficiencies. Shine from the sun, particularly reflecting from the front sight back to the shooter, can be a significant problem. The glare on the front sight can increase the apparent brightness of the light bar on one side of the sight, causing lateral errors in aiming. The simplest solution to the problem of glare is a matte finish on the sights. 'Smoking' a sight by holding a match or cigarette lighter under the sight to deposit a fine layer of soot is a common technique used by many shooters today and countless soldiers in times past. In reducing low-light situations, iron sights sometimes become invisible before the target does. On short-range tactical firearms sometimes the front sight alone, or occasionally the rear sight too, will be marked with a compound such as tritium gas which, being mildly radioactive, makes itself visible by emitting a dim glow. This enables a reasonable aim to be taken in poor light conditions. We will discuss true night vision aids shortly.

Despite their limitations, iron sights can have a place in the armament of a sniper in certain conditions, as may be seen from the following story. Simo Häyhä, the sniper known as 'White Death', was born on 17 December 1905 in Finland, close to the Russian border. At the age of 17 he joined the Finnish militia as a part-time soldier, but remained a farmer and hunter by profession and his farmhouse was full of shooting trophies. Häyhä saw action as a sniper over a period of just 100 days during the 1939–40 'Winter War' between Finland and Russia. In this time he made 505 confirmed kills of Red Army soldiers and a further 37 unconfirmed kills. These were all made with his Finnish militia variant of the Russian Mosin-Nagant called White Guard M/28 'Pystykorva' or 'Spitz'. Besides the Russians he killed with his rifle, Häyhä killed a further 200 men with a Suomi KP/-31 submachine gun, thus bringing his credited kills to at least 705. Unofficially, he is thought to have killed more than 800 men in the 100 days, but his accredited sniping total of 505 is the highest number of confirmed sniping kills in any war.

Häyhä preferred to use the Russian rifle because he said it suited his small frame. He was just 1.6m (5ft 3in tall). More surprisingly, he only used iron sights as opposed to telescopic sights even though they were available. The reasons for this, he said, were that iron sights allowed him to shoot from a lower, less-visible position and did not mist up from his breath in the cold or shine in the sun. Two tricks he used were to compact the snow in front of his position so the muzzle blast did not kick it up and to keep his mouth full of snow so there was no cloud of steam from his breathing. His career ended when he was hit in the jaw by a Russian bullet, but this did not kill him. Häyhä recovered and was made a lieutenant for his efforts. After the war he became a successful moose hunter and hunted with Finnish President Urho Kekkonen.

Faster and more accurate than the iron sight is the reflecting sight. This sighting unit is attached in the place of the rear sight and uses lenses and mirrors to superimpose an image of the target onto a lens with a reticle (crosshairs) or pointer. The way it is constructed causes the image to appear on the reticle only when the rifle is pointing directly at it. Put another way, the image in the reticle is precisely what the rifle is pointing at, whichever angle the shooter views the rear lens of the sight from. Most reflecting sights have zero or small magnification. The British Army issues the SA80 rifle with the SUSAT 4x optical sight as standard issue. The Canadian forces' standard C7 rifle has a 3.4x Elcan C79 optical sight. Both Austria and Australia field variants of the Austrian Steyr AUG, which has had a built-in integral 1.5x reflecting sight since its deployment in the late 1970s.

Telescopic sights, and variants with laser range-finders and computers etc, are the principal sight used by all snipers. A telescopic sight is essentially a telescope with a reticle pattern of some kind in the lenses. The shooter lays the 'crosshairs' onto the target and squeezes the trigger, after allowing for range, altitude, humidity, crosswind and rise or fall to the target. There are many types of telescopic sights on the market, all with their own bundle of features and the technology is improving all the time. For any sort of long-range shooting the telescopic sight is the only choice for a sniper. At ranges of more than a few hundred metres the added accuracy is invaluable and at longer ranges the sniper would not even be able to see the target clearly without them. The only real disadvantages to using a telescopic sight are that it is a little slower to bring to the aim than iron sights and that it is easily knocked out of zero by banging the rifle against something solid. In reply to the speed issue, a properly hidden sniper usually has plenty of time and normally knows where the target will appear. With regard to knocking the sights out of zero, military sight mounts are becoming ever stronger and it is just a fact of life that a sniper must take care of his rifle. Sights also do not look down the inside of the rifle barrel; they are typically set just above it, and the allowance they make for distance to target is a vertical adjustment relative to the vertical set of the weapon. This means that if the shooter does not fire while the weapon is vertically below the sights then the sights will be out of sync with the path of the bullet. Some advanced telescopic sights have a meter and indicator which make sure the rifle is vertical as the shot is taken.

A telescopic sight can be designed with any amount of magnification so the shooter can see a target clearly at any reasonable distance, subject to air clarity, distortion and line of sight. Generally 12x magnification and upwards has been normal for many years and 36x is now commonly employed. As rifles are getting better, and effective ranges are increasing, even more powerful sights are being used and some sights now produce a magnification of more than 40x. The main problem with powerful magnification is that the higher the magnification the greater the 'shake' visible in the sight, just as with powerful binoculars, and the narrower the field of view, so it is sometimes hard to find the target in the sight. To get around the shake problem, a sniper should ensure his rifle is firmly supported for a long-range shot, perhaps using both the front bipod and rear monopod mentioned earlier. Another feature which reduces both shake, and the difficulty in laying onto the target, is that many modern telescopic sights are made with a variable zoom, so the shooter can use just enough magnification for the range he is working at.

The magnification of a telescopic sight is set by the rear lens or ocular. The diameter of the objective or front lens determines how much light can be gathered to form an image. The greater the size of this lens the better the shooter can see in low light. For example, a scope marked 4–12x 50 magnifies from four up to 12 times and has a 50mm (1.9in) light-collecting front lens. The sight's field of view is the width of the visible field. This is usually given as how many metres wide the area is which you can see at 100m (328ft) distance. The greater the field of view, the easier it is to find the target through the scope, but the mechanics of scope production tend to reduce the field of view as the magnification increases.

Another important factor in sight design is known as the exit pupil. This is the width of the image projected back to the shooter's eye from the ocular. It can be calculated by dividing the objective lens diameter by the magnifying power. The ideal is about 7mm (0.28in) as this is the size of the fully dilated human iris in a young man, but this does decrease a little with age. If the zone illuminated by the cone of light beaming from the scope to the eye is more than this then the oversize portion is wasted and the viewed image dimmer than necessary but the eye does not have to be positioned perfectly in line behind the scope. If the exit pupil is smaller than 7mm then the shooter will have to adjust his eye position precisely to centre it on his eye. The more efficient, convenient and comfortable option is a larger exit pupil than the eye requires and a scope which collects more light than strictly necessary.

Eye-relief, already mentioned, is the optimum distance from the rear lens of the scope to the shooter's eye, and is designed into a specific sight. It is the distance at which the image of the target is perfectly focused. Therefore the eye must be held at that focal length to see the image clearly. Put another way, the focal length of the ocular determines the correct eye-relief to be used with that particular sight. The normal focal length is 2.5–10cm (1–4in) and is designed this way to help prevent injury to the eye socket by the recoil of the rifle pushing the sight backwards into the eye socket. The reader will recall that the length of the butt-stock and cheek support may be adjusted to achieve the correct eye-relief.

Before taking a sniping rifle into a combat situation, a sniper must zero the sight so the bullet lands where the sight is pointing. This is done in just the same way as with iron sights, in that the telescopic sight is adjusted to bring the reticle onto the centre of the shot group at an appropriate range. All telescopic sights are fitted with marked dials to make vertical adjustment to set the range or elevation and lateral adjustment for wind or moving targets too. They are also fitted with some type of adjustment, automatic or manual, to bring the image of the target into focus at the sniper's viewing range. Some sights are also fitted with laser range-finders and various types of night-vision equipment (see below).

The reticle pattern on the telescopic sight is very likely a cross with equally spaced markers on each arm. The distance between these markers remains the same size, for a given magnification, whereas the target appears to get smaller. By comparing the apparent size of a man or a vehicle to the reticle markers it is possible to get a good idea of the range. Some reticle patterns have markers for a man-sized target which the shooter can lay against the target and read off the range. Some telescopic sights have a facility to illuminate the reticle pattern for use in poor light. In this case there will be an adjustment control which regulates the brightness level of the lit parts of the crosshairs so that the pattern does not overshadow the target.

Technology can also assist with an accurate shot. There is an integrated ballistic computer/scope system known as BORS, which was developed by the Barrett Firearms Company and which became available around 2007. In essence this BORS module is an electronic bullet drop compensation (BDC) sensor/calculator package intended for long-range sniping out to 2,500m (2,734yd). It works by having the shooter enter the ammunition type into the BORS (using a touch pad on the BORS console) then determine the range (either mechanically or through a laser rangefinder) and twist the elevation knob on the scope until the proper range appears in the BORS display. The BORS then automatically determines the air density, as well as the elevation or depression (shooting uphill or down) of the rifle itself, and incorporates these factors as it works out thexrange elevation allowance required.

There are a number of external factors that can interfere with the accuracy of a telescopic sight. When a barrel has fired a number of rounds in succession it tends to become hot, and in calm air a shimmer may appear above it and distort the view of the scope. In the past some scopes were designed to run the entire length of the barrel to prevent shimmer affecting their line of sight. The modern answer to this problem – one which is becoming almost universal in sniping circles – is to fit a rail along the top of the barrel but thermally isolated from it to reduce the appearance of shimmer.

By featuring glass lenses, a telescopic sight also presents the risk of giving away a sniper's position through reflecting sunlight in a bright flash. To prevent this, the sniper can mount a hood that extends beyond the front of his telescopic sight, casting it into shadow. This hood actually has two purposes: it excludes stray light, which may distort the image, and it prevents reflection from the front lens giving away the position of the shooter. An alternative to the lens hood is a lens filter, which looks something like an expanded metal grating or a honeycomb. With this filter in place, the sight can 'see' well enough through the grating but sun coming from an angle cannot reach the lens to reflect. Various filters are available for fitment to a telescopic sight, such as grey, yellow and polarizing filters to optimize image quality in various lighting conditions and make the target more clearly visible. There are also several types of laser weapon currently being developed for counter-sniper purposes, purposely designed to blind a sniper. The purpose of an eye-safe laser filter, therefore, is to protect the sniper against being wounded or blinded by a laser beam coming from the area of the target, through his sight and into his eye.

Telescopic sights, as issued to snipers, are fitted with covers front and rear which protect the lenses from physical damage by abrasion when they are not in use. These sight lenses are so finely ground that they can be damaged even by contact with the wrong sort of cloth. For this reason, the sniper should take care to keep the protective covers in place when not using the sight and only use the correct cloths provided for cleaning.

Recoil is another physical problem. The recoil from firing a rifle is like a tap from a hammer on the weapon sights. Therefore it is important, particularly with telescopic sights intended for extreme accuracy over long range, that the sights be fastened firmly to the rifle. There have been many mechanisms designed to achieve this, but today by far the most common involves bolting the sight to a 'tactical rail' attached firmly to the barrel or receiver. The idea of a rail to facilitate the attachment of telescopic sights was developed in 1995 by the US weapons research facility at the Picatinny Arsenal as the MIL-STD-1913 rail or STANAG 2324 rail, but it is almost always referred to as the 'Picatinny rail'. These rails are now fitted as standard on many modern weapons, and often retro-fitted to older weapons, so as to allow the firm attachment of not only telescopic sights but also tactical lights, laser aiming modules, night-vision devices, reflex sights, foregrips, bipods and bayonets. Thus equipped, a modern sniper rifle is a very advanced weapon system indeed.

The high-magnification spotting scope used by the spotter in the sniper team is also a sophisticated piece of equipment, and also requires judicious use to be effective. To avoid the shake from which high-magnification equipment always suffers, this spotting scope will be mounted on a tripod of some kind. As with binoculars and scope sights, this device must also be guarded in some way against lens flash. It is probable that in the near future zoom lenses will improve to such an extent that one instrument will become available that performs the jobs of both low-resolution binoculars and high-resolution telescope, and this will save on weight and improve convenience.

Between them, the sniper and the spotter must make an accurate assessment of the range to the target. In the past, before the advent of digital technology, there have been various tools designed to measure the distance to a target by, for instance, focusing a pair of telescopes on the target and using the angle between them to calculate range, or using a telescope with a very short focal length and labelling the focal adjustment ring in such a way that it showed the distance when the target was in focus. Since the invention of laser technology, however, all these systems are obsolete and all sniper teams now use laser range-finders. A laser range-finder works by projecting a narrow laser beam onto the target and then timing the reflection coming back from the point of impact. Even though light travels at 299,000km/sec (186,282 miles/sec) in air, modern technology allows the time it takes for the reflection to come back to be measured accurately and from this the range is calculated to within a couple of metres. Laser range-finders are now built into what look like heavy-duty binoculars and the sniper will also find them built into his sights. Despite the difficult job they do, the operation is pretty much 'point and shoot': the user lays onto the target and presses a button to get an instant readout of the range. The latest type of general military laser range-finders will operate at ranges of just a few metres up to 25km (15½ miles). The more advanced ones are equipped with a digital magnetic compass, inclinometer and computer to provide a magnetic bearing, inclination and the height above sea level of the target. Some of these instruments can even measure a target's speed relative to the observer. When these facilities are combined with a built-in GPS and a digital communications system, then the target information collected by a sniper team can be transferred directly to artillery or air asset fire control computers. (This area of the sniper's work will be considered further in a later section.)

To protect military targets from laser range-finders (and laser-guided weapons) various countries have now developed laser-absorbing paint for their vehicles and clothing. Even without this hi-tech paint, some objects don't reflect laser light very well, so the user must take care and always check that his readings are realistic. The best way to do this is to take a sight on a nearby rock or two and confirm that all the readings are similar.

Obtaining a target and its range is obviously made more problematic at night or in low-light conditions. There are three principal types of device for seeing in the dark: infrared, thermal imaging or an image intensifier.

In terms of infrared, there are two basic forms. Passive infrared devices pick up the infrared spectrum light coming from the moon and stars, and reflected like sunlight from the target, and magnify it. They are not particularly effective aids to night vision, especially on cloudy nights. Active infrared aims a beam of infrared light at the target, which illuminates it to the infrared sensor in the same way that a torch might for a telescope with visible light. Neither a sniper nor any other soldier should use this system, as any enemy equipped with an infrared passive system will see him as plainly as if he were using a torch.

Thermal imaging works by detecting and representing visually the heat given off by the scene under observation, and presents warmer artefacts and human bodies as brighter shades than cooler objects. This equipment does not give the user away to the enemy, as in the case of active infrared. Another advantage of this system is that it highlights exactly what the sniper is looking for – warm bodies and warm vehicles. Because of the different wavelength utilized by thermal imaging equipment, it 'sees' through smoke and rain far better than devices that operate on visible electromagnetic wavelengths. Thus thermal-imaging devices work in fog and smoke as well as in the dark, although they can be quite bulky in nature.

Image intensifiers are the most common form of night-vision device, and probably the best system for actually viewing a scene at night, as opposed to just seeing the warm spots via thermal imaging. They work using the same principle as a TV camera, but are more sensitive to ambient light. They 'see' using the tiny bit of light that is always available outdoors and magnify the brightness enough for the user to be able to see clearly on the darkest night. The early models were awkward to use, as they shut down every time a bright light came 'in shot', but the modern types are superb. All of the above come as helmet-mounted goggles as well as being built into some sophisticated binoculars and scope sights. They are typically accompanied by a hefty price tag.

CHAPTER 3
TAKING A SHOT

In principle, the job of a sniper is quite simple: he must either approach, or await the arrival of, his target without being seen, shoot the individual, and then slip away again. But when the skill sets required are considered in more detail, the task becomes appreciably more difficult. The actual shooting is only a minor part of being a sniper. It is an important part, as it achieves the primary purpose of eliminating the target, but by far the most challenging part of the sniper's job is the field craft element: getting into the right position to take the shot without being seen and then evading the enemy's inevitable response following the shot. If the sniper is seen on the way in to his shooting position, he is either killed quickly, captured or there will be no target in sight. If the sniper does not have his escape planned carefully, any well-organized enemy will kill him with direct or indirect counter-sniper fire within seconds of his taking the shot. (Direct fire, when we are talking about counter-sniper work, denotes a line-of-sight weapon such as a rifle. Indirect fire refers to most air, artillery or mortar fire, when the weapon fires in an indirect arc against the target.)

Weapon Care

All the sniper skills in the world are no use if the soldier suffers a weapon malfunction. Hence one of the fundamental threads of instruction in sniper training is the care and maintenance of the rifle. The first thing an army recruit is taught about a rifle is to strip and reassemble it. Stripping a weapon gets the user familiar with its operation and builds confidence in the rifle, just like learning to pack one's own parachute and knowing why it should open. The rifle stripping drill taught to the soldier is only partial, however, to get at the parts that need regular cleaning and not down to every last nut and bolt. The reason for this is that a military rifle is designed in such a way that a soldier can strip it in the field far enough to clean and lubricate it with minimal or no tools and put it together again without, for instance, a torque wrench. In the case of a sniper rifle, there is an even more specific level of stripping taught for cleaning the weapon and no more, the reason being that stripping may affect the accuracy of the rifle through wear or the spoiling of adjustment. For example, a sniper might well take out the bolt to clean it but he would not, perhaps, remove the telescopic sights from the receiver to avoid spoiling the zeroing. When a rifle needs further stripping than that required for cleaning it is normally a job for the armourer and during the maintenance process the rifle will be re-calibrated and certainly re-zeroed.

Most armies now issue a service log with a sniping rifle and this requires that the armourer enter any work he has done and any defects which he has detected and repaired. The shooter uses the log to record problems, grouping results, number of rounds fired etc. The exact information to be recorded varies slightly between forces and weapons. With a bolt-action rifle, the sniper's daily stripping regime may just be removing the bolt and cleaning it, cleaning the inside of the barrel and wiping down the outside of the weapon. With a gas-operated rifle, the working parts must be removed and the gas piston too as a rule. The exact parts to removed will be taught on the training course relating to the particular rifle.

After stripping and cleaning his rifle every soldier should check for damage to parts and the tightness of screws and adjustors. A sniper should also check the numbers on rifle parts; the number stamped on the receiver, for example, should correspond to the numbers stamped on all the other parts because many sniping rifles are built with specifically matched parts. These parts should not be confused with those from other weapons in the armoury or elsewhere, as mixing parts could lead to stoppages or inaccuracy. (Stoppages are the military name for any mechanical fault in loading, firing, extraction and ejection that stops the weapon operating through its normal cycle.) After reassembly, every soldier should check the working of the action. He will check the smooth return of the magazine slide after depression too, as this can be a cause of stoppages through not lifting the next round to the feed position. A sniper, but not an ordinary soldier, will record faults in the weapon service log book and both will report faults to the armourer.

The sniper will be issued with a special cleaning kit for his rifle. Besides containing the normal infantry issue of a cord to pull a cleaning patch or cloth through the barrel, a wire brush and rifle oil, a sniper's cleaning kit will include special soft cloths for the optics and solvent to remove carbon without using anything abrasive on the barrel or working parts. The purpose of cleaning any rifle is to ensure that it continues to operate efficiently and without stoppages. Dirt or sand get into all mechanical contrivances when they are taken into the field whatever the soldier does to avoid it. Each day that a weapon is out in the field, it should be stripped for cleaning. While stripped the rifle should be wiped down all over with a cloth to remove loose dirt. There may not be much rust, as most modern rifles are made of plastic and rust resistant steel, but it is still vital to inspect the weapon and if any rust is found to remove it and oil the affected part.

The main threat to a rifles smooth and reliable functioning comes from the build-up of carbon residue on the working parts, which is produced by the firing of rounds. With a sniping rifle, the number of rounds fired is far less than one might expect with an assault rifle, and therefore there are fewer carbon issues, but it is also the case that the tolerances in some areas are far tighter, and therefore more sensitive to dirt. With a bolt-action rifle, the barrel, the face of the bolt and the interior of the breech area are most affected by carbon build-up. With a gas-operated weapon, the barrel, the gas piston, the gas regulator (if fitted) and the bolt are most affected. (By their very nature, gas-operated weapons are more prone to excessive carbon build-up than bolt-action rifles.) In the past, cleaning was done with a scraper, a wire brush and then an oily cloth. Today the sniper will most likely be issued with a chemical solvent similar to the RCHS solution. This solution dissolves the carbon without the need for abrasion and wear on the rifle parts. RCHS solution is made from 1L (1.8 pints) drinking water, 200g (7oz) ammonium carbonate and 3–5g (0.1–0.2oz) potassium bichromate (care must be taken to avoid the noxious fumes). The sniper will also be issued with lint-free cloth to avoid leaving fibres in the rifle that might affect its subsequent accuracy.

The barrel of a rifle is normally cleaned with a 'pull through', as mentioned above. This is a length of cord longer than the barrel with a loop at one end and a thin weight at the other. A piece of cloth of a specific size suitable to the calibre is placed in the loop then the cord is threaded into the barrel by means of the weight and the cloth is pulled through behind it. With a clean barrel an oiled cloth is all that is required, but when there may be carbon in the barrel the cloth may be wetted with the cleaning solution. The standard size cloth to fit in the pull through for a 7.62mm barrel is 10 x 5cm (4 x 2in) and comes on a 10cm (4in) wide roll marked into 5cm (2in) rectangles. With most types of barrel steel, a little carbon or dirt is forced into the structure of the barrel interior when the rifle is fired. This can 'sweat' to the surface in the days following shooting, so a rifle which has been fired should be cleaned immediately, subject to operational requirements, then cleaned again a couple of days later even if it has been sitting in the armoury during that time. The sniper should also wipe over the outside of his sights with a clean cloth, but never touch this to the glass as it will become greasy. He may be issued with a special lens cloth for cleaning his sights or he should obtain grease and lint-free cloth for this purpose.

In the field when conditions are dry, the sniper may be concerned about sand sticking to the working parts of the rifle. In this case he should use oil very sparingly, according to his instructions for that type of rifle, as oil is actually more for rust prevention than lubrication. When he is on a mission he will obviously be concerned that the enemy should not identify his shooting position. When a rifle is fired, and there is any trace of oil in the barrel, a plume of smoke is formed by the burnt oil – this is disastrous from the concealment point of view. The way to avoid this is to ensure the barrel is totally dry of oil before firing.

Safety Drills

Alongside maintenance drills, the sniper is also taught how to handle his rifle safely in the field. It is far easier to have an accident with a rifle than a civilian might imagine, particularly when a soldier is under stress or is bored or tired. Therefore all armies have a series of rules for weapon safety, specifically designed to reduce accidents. The following list details the most fundamental safety rules, as followed by all truly professional soldiers, although the sniper would receive further guidance specific to his type of rifle and its safety features.

- Even on operations a rifle must be kept unloaded until leaving base camp. A rifle should be loaded only in a combat or training situation.
- A rifle must be unloaded on return to camp. Most camps have an unloading area where soldiers remove the magazine and any rounds still in the chamber, then pull the trigger while aiming into sand – if a bullet is accidentally fired, it will be safely caught by the sand trap.
- When a soldier is putting a rifle down it should be unloaded so that it will neither go off when knocked nor allow someone to pick up a weapon thinking it is empty and then kill someone else with it accidentally.
- When a soldier picks up a rifle, his own or someone else's, he should check it is unloaded himself. When a soldier hands over a rifle to someone else, he should show the recipient that it is unloaded by opening the breech and 'showing clear', i.e. allowing the other soldier to see inside.
- On a firing range, the shooter should keep the loaded weapon pointing down range at all times. This way, if there is an accidental discharge, the bullet will fly harmlessly away from the group. This rule is important – it seems to be a natural instinct in some people to turn around with a loaded weapon to ask the range officer a question or get a stoppage fixed.
- No soldier should move from the firing point on a range until the range officer has checked all weapons are clear. This way every single person is sure to have cleared his weapon before turning them away from the targets.

Another key element of weapon safety is knowing how to handle a stoppage. All soldiers are taught the stoppage drills appropriate to their rifle, what to do if they pull the trigger and nothing happens. Almost always, for a soldier handling an automatic weapon, the problem will be the round not having fed into the chamber owing to carbon build-up overcoming the gas pressure operating the loading mechanism, or sand in the working parts. Most sniping rifles are bolt-action, therefore a sniper will almost never suffer from a stoppage unless a piece of dirt drops into the working parts from above and stops the breech closing. If there is something stopping the round loading, an experienced sniper will feel the difference in the pressure on his hand as he works the bolt action to load a fresh round. Then he will pull the bolt back and look inside the aperture to investigate. If there is dirt or a misshapen round he will remove this and continue. If he successfully loads a round into a bolt-action rifle, then squeezes the trigger and nothing happens but the click of the firing pin being released, he should work the bolt action to eject the round and reload another, on the grounds that it could be a round with a faulty percussion cap. If another round loads and fails to fire then the firing pin may be broken. A test for this is to inspect the percussion cap on a round which has been in the chamber when the trigger was last squeezed. If it has an indentation in the cap at the centre of its base (but did not fire) then the firing pin is fine and the ammunition batch is suspect. If there is no indentation then the firing pin is either broken or not coming forward and that is a job for an armourer.

Straight Shooting

Though snipers are normally selected from infantry soldiers, the shooting technique required to eliminate a human enemy target at distances greater than a mile is very different to the technique employed by a normal infantry soldier in combat. This is because the latter's purpose is most often to lay down a high rate of suppressing fire to a distance of less than 300m (328yd), combined with the occasional accurate shot to around 100m (109yd). Sniping requires more situational awareness than is possessed by the best competition marksman too, as the marksman does not have to consider the target escaping or shooting back. And killing the enemy at long range requires more field craft and tactical sense than a hunter employs, because deer rarely return accurate fire either.

Of course, any sniper must be able to shoot accurately over long ranges. Sniper trainees are usually expected to be a good shot already with an issue rifle. Some sniper recruits shoot well because they have a natural talent, whereas other people are good shots because they are able to learn the rules of shooting particularly well. The optimum sniper is a naturally good shot whose basic talent is then improved by training. If there is an art to shooting at great distance it is the ability to weigh correctly the many factors affecting the course of the bullet and make the necessary allowances that will cause it to hit the desired target.

Grip and posture are the cornerstones of a good sniper shot. When any shooter is trying to hit a target he should hold his rifle firmly so that the sights can be laid onto the target and held there steadily. The main task of the shooter's hands is to pull the butt of the rifle back into his shoulder. Besides reducing the impact of recoil, a firm grip holds the butt steadily in the crook of the sniper's shoulder and therefore allows consistently accurate shooting through steadiness and repeatability. The shooter's trigger hand should be around the pistol grip, or a narrowed portion of the stock, pulling backwards. His non-trigger hand should be under the fore-end and pulling back too. In the case of a sniping rifle fitted with a bipod, the sniper is always trying to use the bipod under the barrel as the front support. Here the task of the non-trigger hand is not normally support and it may be placed, as with a machine gun, on the butt-stock in front of the face. When the sniper cannot use the bipod, or doesn't have one fitted to his rifle, then he should first try to rest the fore-end on a solid support (sometimes axsmall sandbag is carried around for this purpose) and failing that support the barrel with his non-trigger hand gripping in the same place as the bipod. When supporting the fore-stock with his hand, every shooter should keep his forearm vertical on the lateral plane to avoid a tension-induced wobble, which will open up his group into a diagonal arc across the target centred on his supporting elbow.

Another important consideration is 'cheek weld'. This term refers to the way the sniper's cheek lies against the butt of the rifle. By correct adjustment of the butt and the cheek support on his rifle, a sniper should be able to maintain exactly the same comfortable cheek position on the butt for each shot. He should do this because cheek weld affects his eye position vertically and horizontally relative to the sight, and hence very slightly alters the appearance of the image of the target in the sight field, thus affecting the consistency of his shots.

After learning the basic grip of the rifle, the sniper is taught four basic shooting positions: prone (lying down), sitting (cross legged on the floor), kneeling (on one knee) and standing. Each of these positions has its place in normal infantry training, but they are somewhat different when using a support or bipod. The sniper will lie prone when on flat ground, but sit, kneel or stand to get his height correct behind a rifle resting on its bipod, which in turn is on, for example, a sandbag, wall or parapet.

Looking in more detail at the shooting positions, when the sniper is prone his leg on the same side as the trigger finger should be in line with the barrel and the other leg splayed a little to the side for comfort and lateral stability. This is the most stable, and therefore accurate, position for shooting and should be employed whenever possible for this reason. It also has the advantage that it offers a small visible target to the enemy. When adopting the sitting position, many snipers like to cross their ankles and rest their arms just above the elbow on their knees (not exactly on the elbow, as this point-to-point contact is not steady). This position might be used by a sniper when firing from grass that is too long to see the target from a lying position, and there is nothing to rest on or hide behind.

The kneeling position is a rather awkward shooting stance, and should be used only when it is required to take a higher position than sitting, such as when firing out of a low window or using a low table for support. The sniper should lean forward onto the rest if possible, to improve steadiness. Standing is used where the sniper has to stand to fire through a higher aperture. When adopting this posture, he should try to keep his back straight to avoid tension and shake. Spreading his feet side to side will give a steadier aim too.

These four main positions from which the sniper can shoot his rifle are all intended to help him to hold the rifle as steady as possible in the circumstances. To achieve maximum steadiness he should seek to avoid muscle tension. One way in which he can do this is as follows: he takes up the shooting position and gets comfortable then looks through the sights. The target will most likely not be in the centre of his sight picture. If this is the case then rather than twisting his body to move the rifle and sights onto the target he should adjust where he is lying or kneeling until the target appears naturally and comfortably in the centre of his sights.

Once a sniper is in the correct shooting position, and has identified a viable target, he is now ready to pull the trigger and take his first shot. The important thing about pulling the trigger is that a shooter must avoid yanking the trigger back abruptly with his finger, an action that will pull the rifle off aim at the last moment. Instead he must squeeze the trigger by contracting the muscles of his forearm to cause the thumb and forefinger to come towards one another. This movement enables a steady, controlled increase in pressure on the trigger without pulling the weapon off aim. It matters which part of the finger rests on the trigger. The shooter has more control if he uses the second pad of his trigger finger. Because this shortens the effective lever against the trigger, the shooter has more strength and therefore a smoother action.

Breathing and timing are also critical factors in the steady shot, as the US Army *Sniper Training* manual makes clear:

Breath control is important with respect to the aiming process. If the sniper breathes while trying to aim, the rise and fall of his chest causes the rifle to move. He must, therefore, accomplish sight alignment during breathing. To do this, he first inhales then exhales normally and stops at the moment of natural respiratory pause.

a. A respiratory cycle lasts 4 to 5 seconds. Inhalation and exhalation require only about 2 seconds. Thus, between each respiratory cycle there is a pause of 2 to 3 seconds. This pause can be extended to 10 seconds without any special effort or unpleasant sensations. The sniper should shoot during this pause when his breathing muscles relax. This avoids strain on his diaphragm.

b. A sniper should assume his firing position and breathe naturally until his hold begins to settle. Many snipers then take a slightly deeper breath, exhale, and pause, expecting to fire the shot during the pause. If the hold does not settle enough to allow the shot to be fired, the sniper resumes normal breathing and repeats the process.

c. The respiratory pause should never feel unnatural. If it is too long, the body suffers from oxygen deficiency and sends out signals to resume breathing. These signals produce involuntary movements in the diaphragm and interfere with the sniper's ability to concentrate. About 8 to 10 seconds is the maximum safe period for the respiratory pause. During multiple, rapid engagements, the breathing cycle should be forced through a rapid, shallow cycle between shots instead of trying to hold the breath or breathing. Firing should be accomplished at the forced respiratory pause.

– US Army FM 23-10, 3-3 (1997)

In an ideal world a sniper should squeeze the trigger when his breathing is half exhaled, so that the body is under least tension. Holding a full breath or shooting from maximum outbreath is never advised, as the muscular tension of the former and the oxygen depletion of the latter can both induce shake. The real world, however, makes it more difficult to find the right moment in the breathing cycle, because the sniper's spotter might be telling him when to fire as the wind drops or the target moves into an ideal position. This is one reason why sniper and spotter have to know each other well enough to understand each other.

Some snipers have a tendency to rush their shots and move as soon as they have fired, especially if they are in a hurry to get away from the firing position to avoid being shot, shelled or captured. This can lead to a tendency to move or 'snatch' the rifle while still *about to* fire. To counteract this tendency, the sniper should habitually count to one or two after pulling the trigger, not moving at all during this period, and then he will not snatch his rifle off aim at the last moment.

The basis of all good shooting is being able to put a tight group onto a target consistently when aiming at a given mark. The smaller the group at a given range the better the shooting. If a shooter can produce a small group then all he has to do to hit a target, living or dead, reliably is adjust the sights so that the group he fires is centred on the point he wants to hit. This process of bringing the centre of the group onto the target by adjusting the sights or scope is called zeroing. The group shot by different people with the same rifle and sights set the same way, however, lands in different places owing to their differing shooting techniques and personal idiosyncrasies. Therefore a rifle has to be zeroed for one individual user if it is going to be used for accurate work.

When a scope is first bolted to a rifle it will not be pointing exactly where the bullet is going to land. So first of all the scope itself needs to be lined up roughly with the bore of the rifle barrel; there is a tool that fits into the barrel which allows this to be performed quickly and easily. Once the sights are roughly lined up with the barrel and set to 'no wind offset', the sniper goes onto a windless range and shoots ten rounds or so at the centre of a target, at a distance appropriate to the rifle, and with the sights set to that distance. A 2.54cm (1in) white square patch at 274–548m (300–600yd) is normal for sniper zeroing, whereas an infantry rifle would be zeroed at 91m (100yd). The sniper ought to lay down a fairly tight group, perhaps 5–10cm (2–4in) across depending on the rifle. The centre of his group being a few inches high or low, left or right of the white patch, does not matter so long as it is on the target board and therefore the offset is measurable. Then the sights are adjusted so that the next group he shoots is on the centre of the target. If the group is not quite on target, then further adjustment is made and more groups shot until it is.

In theory, once the sights are zeroed and the sniper shoots with correct technique the result ought to be a direct hit on the target. Yet we have already touched upon some of the factors that affect the travel of a bullet once it has left the barrel, and the sniper must make allowances for all these effects. Up to a few hundred metres of range, wind and target movement are the only significant factors to allow for, so the sniper can just set the scope to take these into account and put the centre of the crosshairs onto the target. As the range increases, however, some other factors begin to make themselves felt. In the following section we will consider the main factors affecting the bullet's path at long ranges, beginning with the most significant and ending with those of a more subtle nature.

In principle the sniper, or his spotter, will estimate the factors that have significance for a particular shot and adjust the scope to compensate for their effect. The following will give an idea of what a sniper or his spotter has to consider when aiming at a distant human target.

Allowing for drop is a central challenge in the accurate long-range sniper shot. As soon as a bullet leaves the barrel of a rifle it begins to fall, with the speed of fall increasing over time at a rate of 9.8m/sec (32ft/sec). What this means is that at a short range the bullet seems to travel flat, because the bullet covers the space so quickly, but at longer ranges the shooter must allow for the fall of the bullet. A bullet leaving the muzzle at 1,036m/sec (3,400ft/sec) begins both slowing and falling straight away. Therefore in the first second it might travel 975m (3,200ft) but will definitely fall 9.8m (32ft) below the point the shooter would see looking down the horizontal bore of the rifle. In the next second it will travel, say, another 853m (2,800ft) – the round is continually slowing – and drop another 19.5m (64ft), making 29m (96ft) dropped. In the third second it will travel another 671m (2,200ft) and drop another 29m (96ft) making 60m (196ft) dropped in total.

These speeds are a rough estimation and depend on the aerodynamics of the bullet, bullet weight etc. Every type of bullet is not only different to every other, but decelerates at a different rate according to its initial velocity, air temperature and pressure and other factors. But the drop is a matter of physics and precise. A NATO 7.62mm bullet loses two-thirds of its velocity and nine-tenths of its kinetic energy in 1,400m (4,593ft), but a specialized sniping bullet is more aerodynamic and has a higher mass/drag ratio, so it loses speed more slowly due to more effective aerodynamic flight characteristics. To take the fall into account and still have the bullet strike on target, the sniper has to adjust the sights for range. Effectively, the sights are set to look lower relative to the barrel when shooting at targets further away, hence when the gun is fired the bullet travels in an upward rising arc to the target to allow for drop. Different barrel temperatures, bullet weights and propellant loadings alter the bullet's muzzle velocity, and therefore drop over a given distance, so the allowance which must be made for distance changes when using different ammunition. On top of these considerations there are atmospheric conditions to take into account. At higher altitudes the air is thinner so it slows the bullet down less and therefore a rifle shoots 'high'. When the temperature is lower the air is thicker and therefore slows the bullet more and a rifle shoots low. Humidity, mist and rain also slow bullets down.

Alongside drop the sniper will also have to allow for gradient; when he is shooting at someone higher than himself he is shooting uphill, so the bullet will lose energy faster – it requires energy to lift a bullet uphill and so the bullet will decelerate faster – and strike low relative to the normal sight settings for that range. The converse is of course true for shooting downhill. All snipers are issued with either tables or computers that allow them to calculate how much lower their aiming point should be set for an uphill target at whatever range and the converse for a downhill.

Then there is the matter of wind to take into account. Wind blowing from a shooter's left or right pushes the bullet away from where the zeroed scope is pointing and in the same direction as the wind is blowing; so the sniper must compensate by aiming-off towards the wind's source if he wishes the bullet to land on target. A wind of 48km/h (30mph) blowing from 90 degrees left of the line of shooting, for example, will deflect a NATO 7.62mm bullet by 1.67m (5ft 6in) at a range of 457m (500yd). But at 914m (1,000yd) it will deflect the same bullet by 5.7m (18ft 8in) because the pressure is cumulative while the bullet is also slowing down.

Every sniper team is equipped with tables and calculators to make allowance for the huge wind deflection that occurs at long range. Effectively, in the case above, the scope must be adjusted so that when the crosshairs are on the target at 914m (1,000yd), the barrel is actually pointing 5.7m (18ft 8in) to the left of the target. If the wind is behind the shooter, it makes the bullet travel slightly high and if the wind is coming from the front then the bullet will travel low. Counter intuitively, if the wind is on the sniper's quarter – that is blowing at 45 degrees across the line of sight – the sniper must allow three-quarters of the deflection for the 90-degree wind at whatever speed, rather than the half allowance that a layperson might expect.

Of course the effect of wind varies on a sliding scale depending on its strength and its direction, the weight of the bullet, the bullet velocity and the distance from the muzzle at which the wind has its effect. Such factors vary for different bullets and propellant loads, so there are different tables and ballistic computer settings for each and every combination of rifle and ammunition. With older scopes the sniper would set the range with a dial and then aim off 'manually' by laying one of the equally spaced marks running along the lateral line of the crosshairs onto the target. If the wind was coming from the right he would lay the first or second mark left of centre on the reticle's lateral line onto the target. (The second mark would move the barrel more to the right than the first mark and so on.) Scopes are getting more capable all the time, however, and more and more factors such as range, wind, elevation and so forth are being dealt with either by settings on the scope or by computers that adjust the scope automatically when it is laid onto the target. On the latest models, the range allowance is made by a built-in laser range-finder calculating the range to the target and adjusting the scope itself. When shooting up or down hill, some scopes assess how far from level the rifle is being held and combine that with the range to make allowance. There are even some scopes that allow for the wind over range automatically, by combining a given wind speed with the range.

Clearly, over a distance of more than a kilometre, and with several seconds of bullet flight time, the wind can vary in speed and direction. At extended ranges the sniper may have to aim off by many metres from the target with a steady wind and when this wind varies the necessary allowance will change quite dramatically too. All this estimation makes shooting at long range with a variable wind very difficult, so the ideal technique is to set the scope for either a certain wind speed, or none at all, and have the spotter call out when those conditions pertain and the sniper can usefully fire. The spotter will assess the wind by watching the effect on foliage, smoke, dust or whatever is available.

Another complication is that if it takes the sniper's bullet three seconds to travel 2,500m (2,734yd) to a target, then the target might have moved by the time the bullet gets there. If the target is walking left to right he might have walked 2.5–5m (4–6yd) from the moment the trigger is pulled to the point the bullet reaches his original position. The sniper team has to estimate where the target will be when the bullet reaches him and aim off accordingly on top of any range, wind and other allowances. By far the best plan is to attack a stationary target that appears intent on remaining still. Of course this is not always simple to judge, unless the target individual is engaged in a task that keeps him in one place. If a moving target must be attacked, then the sniper should attempt to shoot when he is coming towards or moving away from the sniper, as this will eliminate the aim-off component for movement. If the target is walking at an angle, the sniper must allow the 'apparent' distance the target will walk over the bullet flight time, then aim ahead of the target by the apparent distance he will have moved to the sniper's left or right side.

Temperature and humidity also affect the bullet's flight to target. Warm air expands and therefore becomes thinner, which means that a bullet travels through it more easily, slows down less and effectively flies flatter. There is an allowance table telling the sniper how much lower to aim as the air gets warmer. Counter intuitively, a bullet flies very slightly higher in humid air, but lower in rain, and again there are tables supplied to the sniper to make allowance for this fact. Air pressure makes a difference too – a low-pressure air mass makes a bullet travel slightly high, unless the low-pressure system has actually brought rain, in which case the bullet will travel low.

Although the gyroscopic spin applied to a bullet by rifling doubtless enhances the accuracy and range of a sniper rifle, over long ranges it introduces its own problems for the sniper. The difference between the long axis of the bullet upon which it is spinning and the direction of the velocity of the centre of gravity, caused by the path of the trajectory, is called the 'equilibrium yaw' or 'yaw of repose', or more simply 'gyroscopic drift'. For reasons of physics this generates a tendency to movement at 90 degrees to the direction of travel. Put simply a bullet spinning clockwise will tend to move right and a little high of its expected path. At 914m (1,000yd) a US military M193 ball round moves 58cm (23in) to the right through gyroscopic drift, but sniper ammunition tends to be deflected less than this. There are tables issued to snipers that make this calculation, or it is made automatically by the scope computer.

Spin-stabilized projectiles are also affected by something called the Magnus Effect. All this means is that the spin of the bullet creates a force acting either up or down at 90 degrees to the sideways pressure of the wind. If the wind is level from the right and the bullet is spinning clockwise then the bullet is pressed down. Although this effect is quite small it acts on the centre of pressure of the bullet and, because this is generally ahead of the centre of gravity in a normal bullet design, it tends to destabilize the bullet. When the centre of pressure is behind the centre of gravity, however, the Magnus Effect tends to aid stability. To make the Magnus Effect more interesting, it actually acts on the centre of pressure apparent in the jacket of air travelling around the bullet, as opposed to that created by the bullet jacket design, and the shape of this air jacket changes as the bullet's velocity reduces. The resultant yawing effect is more marked in low-drag bullets because of their elongated shape.

Added to the Magnus Effect is the Poisson Effect. This effect relies for its occurrence on the nose of the spinning projectile being above the line of its trajectory, as it must always be. Therefore a cushion of air builds up under the nose of the bullet and the increased pressure there leads to drift in the direction of the contact. A bullet spinning clockwise will drift slightly to the right.

The list of considerations for the sniper extends with the Coriolis Effect. When a sniper shoots a rifle he is sitting on a moving sphere – the Earth – the target likewise, but some distance away. The bullet has inertia and is moving in a curved path through a moving gravitational field to a moving target. Gravity acts on the bullet with the effect of making a bullet fired in the northern hemisphere appear to move to the right with reference to the rifle and the target. Actually the Earth is moving the other way. In the southern hemisphere, the opposite appears to happen and the effect is far stronger at the poles than theequator so the shooter would need to know his latitude to work out the effect. In reality, this effect is more significant for ballistic missiles than rifle bullets owing to their longer flight time.

The Eötvös Effect is all about centrifugal force and is caused by the spin of the Earth acting on the bullet as it travels laterally on the surface east–west or the reverse. The physics are complicated, but the effect is that a bullet fired to the east travels low and a bullet fired west travels high. Of course, the amount of deflection is relatively small with rifles as opposed to artillery.

Coriolis Drift is the sum total of the Coriolis Effect and the Eötvös Effect. Both of these effects are features of the way gravity acts on objects moving through its field as opposed to the atmospheric effects which we have looked at previously.

Thankfully, modern snipers have access to a range of technologies to help them automate the calculations required by the above effects.

Yet the traditional skills and brainwork also have to be reinforced. A special forces sniper team may very easily be in a situation where they cannot be resupplied when electronic equipment goes defective, and have to make do with the basics to get the job done. A range card, for example, is simply a card or paper with the shooting position marked on one edge by a cross. Radiating out from this cross are concentric semicircles representing distances of, say, 100m from the shooting position. Features and areas where targets are expected may be marked on this card by the spotter or the sniper to save time finding the range when a target shows itself. Such is the range of the modern sniping rifle that an ordnance survey map could be used as a range card in suitable circumstances. Similarly, a sniper's log or target log is somewhere between an old-fashioned almanac giving the phases of the moon and a diary. On the almanac side the log gives the user all manner of useful information relating to how this rifle shoots in different conditions. This information allows a sniper to make or check his calculations for a first shot. On the diary side the log is used to record the strike of rounds fired in different conditions. By doing this regularly, the sniper builds up a picture of how his rifle fires in different winds and at different ranges. The logs are therefore a great aid to making a one-shot kill when he comes up against similar conditions again.

The US Army *Sniper Training* manual summarizes what it calls the 'integrated act of firing', a system introducing consistency into the process of taking a sniper shot. The clarity of this material is such that it is here quoted at length:

Once the sniper has been taught the fundamentals of marksmanship, his primary concern is his ability to apply it in the performance of his mission. An effective method of applying [the] fundamentals is through the use of the integrated act of firing one round. The integrated act is a logical, step-by-step development of fundamentals whereby the sniper can develop habits that enable him to fire each shot the same way. The integrated act of firing can be divided into four distinct phases:

a. **Preparation Phase.** Before departing the preparation area, the sniper ensures that –

(1) The team is mentally conditioned and knows what mission

they are to accomplish.

(2) A systematic check is made of equipment for completeness and serviceability including, but not limited to –

(a) Properly cleaned and lubricated rifles.

(b) Properly mounted and torqued scopes.

(c) Zero-sighted systems and recorded data in the sniper data book.

(d) Study of the weather conditions to determine their possible effects on the team's performance of the mission.

b. **Before-Firing Phase.** On arrival at the mission site, the team exercises care in selecting positions. The sniper ensures the selected positions support the mission. During this phase, the sniper –

(1) Maintains strict adherence to the fundamentals of position. He ensures that the firing position is as relaxed as possible, making the most of available external support. He also makes sure the support is stable, conforms to the position, and allows a correct, natural point of aim for each designated area or target.

(2) Once in position, removes the scope covers and checks the field(s) of fire, making any needed corrections to ensure clear, unobstructed firing lanes.

(3) Makes dry firing and natural point of aim checks.

(4) Double-checks ammunition for serviceability and completes final magazine loading.

(5) Notifies the observer he is ready to engage targets. The observer must be constantly aware of weather conditions that may affect the accuracy of the shots. He must also stay ahead of the tactical situation.

c. **Firing Phase.** Upon detection, or if directed to a suitable target, the sniper makes appropriate sight changes, aims, and tells the observer he is ready to fire. The observer then gives the needed windage and observes the target. To fire the rifle, the sniper should remember the key word, 'BRAS.' Each letter is explained as follows:

(1) *Breathe.* The sniper inhales and exhales to the natural respiratory pause. He checks for consistent head placement and stock weld. He ensures eye-relief is correct (full field of view through the scope; no shadows present). At the same time, he begins aligning the crosshairs or front blade with the target at the desired point of aim.

(2) *Relax.* As the sniper exhales, he relaxes as many muscles as

possible, while maintaining control of the weapon and position.

(3) *Aim.* If the sniper has a good, natural point of aim, the rifle points at the desired target during the respiratory pause. If the aim is off, the sniper should make a slight adjustment to acquire the desired point of aim. He avoids 'muscling' the weapon toward the aiming point.

(4) *Squeeze.* As long as the sight picture is satisfactory, the sniper squeezes the trigger. The pressure applied to the trigger must be straight to the rear without disturbing the lay of the rifle or the desired point of aim.

d. **After-Firing Phase.** The sniper must analyze his performance. If the shot impacted at the desired spot (a target hit), it may be assumed the integrated act of firing one round was correctly followed. If however, the shot was off call, the sniper and observer must check for possible errors.

(1) Failure to follow the keyword, BRAS (partial field of view, breath held incorrectly, trigger jerked, rifle muscled into position, and so on).

(2) Target improperly ranged with scope (causing high or low shots).

(3) Incorrectly compensated for wind (causing right or left shots).

(4) Possible weapon/ammunition malfunction (used only as a last resort when no other errors are detected).

Once the probable reasons for an off-call shot [are] determined the sniper must make note of the errors. He pays close attention to the problem areas to increase the accuracy of future shots.

– US Army FM 23-10, *Sniping Training*, 3-7 (1997)

Working with a Spotter

Normally all snipers work as part of a two-person team: a sniper and a spotter. This proven system has been in place since World War II, and will probably continue to be used for as long as snipers have a place in armies. The spotter's job is quite simply to find targets for the sniper, help with assessing the range, keep a look-out for anti-sniper teams and reduce the time the sniper has to spend straining his eyes watching a target area.

A theoretical example will illustrate the cooperative nature of the sniper team. A sniper team have taken up position on a village roof top, well hidden of course, and they are looking out across flat, fairly open farmland for targets of opportunity, such as insurgents setting up mortars or an ambush, or laying an IED. The sniping team have to assume that targets may appear anywhere in their field of view – to left or right out to the extreme range of their rifle. Given a random spread of targets, it is likely the targets that do present will not be within a couple of hundred yards because of the way vision works. Consider this. Suppose the sniper's field of view is 180 degrees from left to right. The area in front of him and within 200m (219yd) is an area of roughly 6 hectares (16 acres). However, the sniper can hit a target out to 2,400m (2,625yd), and is therefore controlling an area of 942 hectares (2,260 acres) to his front. This is far too great a space to be controlled by one man, hence he needs to be accompanied by a spotter with a pair of wide-angle binoculars and a high-power spotting scope. The scope in itself is not much good for scanning the whole area quickly to find a new target, whereas the spotter's binoculars have a much wider field of view. The spotter doesn't need to ID the target's cap badge at a range of a mile; he needs to scan the whole area and spot someone who looks as if he is potential target. Then he checks with his high-magnification spotting scope before he tells the sniper where the target is and the sniper can take a look through his own scope.

Apart from simply multiplying the pairs of eyes on the battlefield, there are further benefits to the sniper/spotter relationship. To begin with, the sniper concentrating down his scope is a very easy target for someone approaching from an unexpected direction, so the spotter carries an assault rifle to protect the sniper. There are many missions where the target area needs to be watched closely for hours or days so the sniper and spotter can take it in turns to do that job while the other man rests his eyes. It goes without saying that creeping behind enemy lines or sitting up in a hide on a tall building for a week or two is a lonely job, and having a partner makes the work less lonely.

Then there is the actual shooting. When the team are considering shooting a target, the sniper will be watching the target or area and the spotter will be doing a whole range of other jobs. First of all there are the calculations of range, wind, altitude, humidity and target aim-off to work out. It is normal for the spotter to be equipped with a notepad and a laptop computer specifically for performing these calculations. Even if he is sometimes working as a backup to the high-tech scope which can work these things out for itself, he is there if it goes wrong. And besides looking for fresh targets, the spotter will be checking around the team's position for anti-sniper teams stalking his team. Depending on the type of mission and situation the spotter may also be in radio communications taking orders as to the priority of engaging the target or even if the target should be engaged at all. On tasks where both the sniper and spotter may need to take a shot the spotter may bring his own sniping rifle, zeroed for himself.

Sniper Training provides a useful summary of the distinct roles of sniper and spotter:

Each member of the sniper team has specific responsibilities. Only through repeated practice can the team begin to function properly. Responsibilities of team members are as follows:

The sniper –
- Builds a steady, comfortable position.
- Locates and identifies the designated target.
- Estimates the range to the target.
- Dials in the proper elevation and windage to engage the target.
- Notifies the observer of readiness to fire.
- Takes aim at the designated target.
- Controls breathing at natural respiratory pause.
- Executes proper trigger control.
- Follows through.
- Makes an accurate and timely shot call.
- Prepares to fire subsequent shots, if necessary.

The observer –
- Properly positions himself.
- Selects an appropriate target.
- Assists in range estimation.
- Calculates the effect of existing weather conditions on ballistics.
- Reports sight adjustment data to the sniper.
- Uses the M49 observation telescope for shot observation.
- Critiques performance.

– US Army, FM 23-10, *Sniping Training*, 1-4 (1997)

Considering all these tasks to be performed by the spotter, it is not surprising that in some armies on some missions a third member is added to the team and called a 'flanker'. His task is to be responsible for the defence of the position, particularly the rear, and leave the spotter more time to concentrate on the technical jobs. In a normal shooting situation, if there can be such a thing, the sniper will call out 'ready' when he is sighted on the target and ready to take the shot. The spotter will be waiting for this and watching the wind and other effects. When he judges the right moment, he will call 'shoot' or make some other pre-arranged signal and the sniper will then fire. When the shot is finally taken the spotter will watch for the shot impact and call a correction to the sniper if there was a miss and another shot can be taken. (Note that a bullet leaves a trail of disturbed air that can be observed from behind, and the strike of the bullet will sometimes be visible through a telescope when it throws up dust or knocks a chunk out of a wall.)

On any mission where there are enemy to return fire, a sniper should try to avoid shooting when it is dark, and if he has to shoot then he should move immediately afterwards. All secrecy is lost if a shot is taken at night because of the muzzle flash, which will light up the area and give away the team's position. Of course, in some situations, such as a hostage-release and firing from a known sangar type position within a base camp, firing at night has fewer disadvantages. But when he is performing the classic sniper mission of stalking the enemy behind their lines, then shooting at night should be avoided whenever possible.

In most sniping situations, the sniper will usually take a shot as soon as a target presents itself. There are several situations, however, when a sniper may delay engaging. One is where the team are working in a hostage-rescue or special ops raid situation and waiting for the 'shoot' command from higher up. In this instance, the sniper or spotter may be in constant communication with the commander or the other snipers, as shots may have to be sanctioned by a politician or senior rank. There might also be the need for several snipers to hit their targets simultaneously, hence the moment of shot is decided centrally. In another situation, the target may be on the edge of the sniper's range and the weather may be poor. In this case the sniper may choose to call in indirect fire from mortars, artillery or aircraft to be sure of a kill.

Observation & Fire Control

Because the sniper team's role includes a close and steady observation of the enemy, it is natural that the team is put to uses other than just sniping. It has useful reconnaissance and intelligence-gathering applications, reporting back from a hide position about enemy movements and strengths. More aggressively, the team can also act in a fire-control capacity, acting as forward observers to guide artillery fire or an air strike onto the enemy position.

The problem for artillery batteries and high-altitude or distant strike aircraft is that they often cannot see what they are shooting at, or need someone on the ground to identify targets. The way they deal with this is to deploy an officer called, by the British, a fire control officer (FCO). His job is to go up to the frontline, or wherever the enemy are, and look for targets. When he has a target for, say, an artillery battery he calls back to them by radio and speaks to the fire direction centre (FDC) at the battery giving them the target. He has a set format for this procedure, which includes information about the map grid reference of the target, how long the target may stay in place and what ammunition to use against it. The FDC, generally sited at the battery itself, then works out the bearing and range for the guns and gives them their orders as to what ammunition to use, be it air burst, ground burst, phosphorous etc. From this point different armies do things differently. What used to happen is one gun would fire a ranging shot and that would be corrected onto the target by the FCO before the rest of the battery opened up and destroyed it. (Today, with some units, the equipment is so accurate they expect to hit the target first time and 'fire for effect' straight away so as not to warn the enemy with the ranging shot.) If the target was moving, perhaps a convoy, then the first salvo would require correcting as the convoy moved on. This last technique is called 'walking fire'. Whatever the situation, once the guns were on target they would then continue to fire until the FCO let them know that it was neutralized.

In the future, all snipers will be trained to call in fire from whatever indirect-fire units are available. This is a trend developing around the world and already implemented in most Western armies. The step is in many ways a natural one. The job of an FCO, or any special forces artillery spotter, is very much like that of a sniper – he has to get in sight of important enemy targets and possibly stay there for a long time, all the while remaining hidden from view. The US Army was the first to begin to train its sniper teams in the art of artillery fire control. The principles are simple enough for anyone who can pass a sniper course; the only consideration is that the fire-control orders vary according to what fire is being called in.

Taking on the role of fire control has proven very popular with sniper trainees, as calling in artillery is quite a buzz, with its accompanying roar and devastation. Not for nothing did Joseph Stalin, dictator of the Soviet Union, call artillery 'The God of War'. The author was once shelled by 155mm artillery and it felt like the end of the world. Some sniper missions, such as assassinations, are going to remain the province of rifles only, but as far as possible a sniper should always have the option of calling in indirect fire to destroy a target he has found. He might come across a whole crew of senior officers or something else of great value merely by chance, and upon such opportunities turn the outcome of battles and wars, providing the opportunity is acted upon. And there is one major advantage for the sniper calling in indirect fire: it does not give away his position.

In the time before digital electronics, a person wishing to call in fire of any kind would know his own position and from that work out, by bearing and range-finder, the position of the enemy target. This can still be done if there is a technical malfunction, but today the process is much quicker and easier because there has been developed what the military call a laser target designator (LTD). An LTD looks something like a large, heavy-duty pair of binoculars. All the user has to do is look through the viewer at the target and line up the reticle onto the target then press one button to select type of fire and another to send the message to artillery or whatever is providing the fire. Some early models required the operator to radio in the type and duration of fire from the grid reference produced by the equipment, but this is not now common. The message the LTD sends to the guns is the exact position of the enemy worked out by knowing its own position and altitude through GPS, getting the range of the enemy with a laser, their bearing with a compass and their altitude with an inclinometer and computer. It works the information out then sends the message to the artillery or air fire-control computer and its handlers at the press of a button. The fire-control computers then work out the actual bearings for their guns and do the job. At the time of writing, computers are beginning to take up all the work between LTD and firing shells. This is so quick, clean, efficient and lacking in opportunity for error that it will soon replace all other forms of fire control. With an LTD, and enough artillery fire or strike aircraft on tap, one man could destroy an army.

Before a demonstration of counter-sniper tactics for use in no-man's land, a scout is photographed helping his observer into camouflage gear. *(Getty Images)*

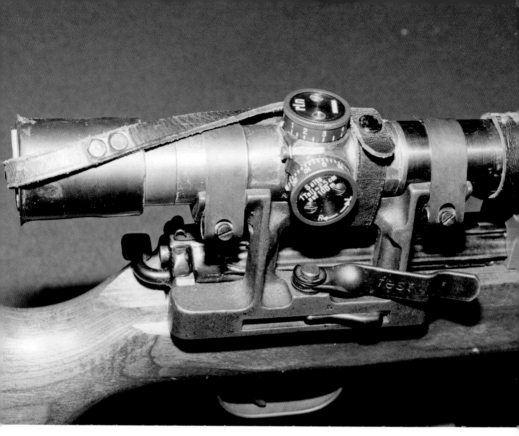

Shown here is Germany's own version of the Russian scope shown previously; this is the Zf4 (4x magnification), pictured mounted on a Mauser 98K rifle. *(Courtesy of Martin Pegler)*

The Enfield Mk 4 (T) shown with the No. 32 scope and mount. Although heavy, the scope was extremely hardy and saw use into the post-war period. *(Courtesy of Martin Pegler)*

Workers in a factory set rifle barrels by machine during World War II. *(Corbis)*

Military rifle
ammunition used from
the beginning of the
20th century. From left:
US .30-06in, British
Mk VII .303in, German
7.92mm Mauser,
Soviet 7.62 x 54mm,
French 8mm Lebel
and Japanese 6.5mm.
(Courtesy of Martin Pegler)

A US Marine is pictured posing with the new M1903-A4 sniping rifle. The low power (2.5x magnification) M73B1 scope ensured it was never suitable for long-range sniping. *(USMC)*

The Pacific during World War II. A Marine is photographed cleaning his Springfield rifle during a lull in the fighting. In World War II the rifle's enemy was rust owing to the inferior metals used in manufacture, whereas today the main problem is carbon build-up. *(USMC)*

An M3 active-infrared Sniperscope for night vision mounted on an M1 Carbine during the Korean War. As can be imagined, the sheer size and weight of the scope meant that soldiers were reluctant to carry it far, but the ability to see at night proved its worth so the M3 saw service in Vietnam. *(US Army)*

Parts of the Israeli Galil assault rifle are shown here by a worker at the Colombian Military Industry organization, which manufactures these weapons under licence. *(Getty Images)*

A US recon team pull a suspected Vietnamese sniper from a hole near Saigon during the Vietnam War, 1968. *(Topfoto)*

A Russian Dragunov SVD sniper rifle of 1986 vintage fitted with PSO-1 4x 24 scope. Based on the design of the AK-47, this rifle was adopted in the mid-1960s as the designated marksman/sniping rifle of the Eastern Bloc powers. *(Topfoto)*

The Barrett M82A1 .50-cal sniper rifle photographed in Kunduz, Afghanistan. This is one of the later designs of sniper rifle developed to fire the Browning Machine Gun (BMG) round, and its derivatives, for maximum hitting power. Note the muzzle brake to reduce the severe recoil. *(Corbis)*

A US Marine sniper aims a 7.62mm M40 bolt-action sniper rifle during mountain warfare training in the Californian Sierra Nevada mountains. *(Corbis)*

British Army AWM (Arctic Warfare Magnum) L115A3 bolt-action sniper rifle chambered for the .338in Lapua Magnum. This same rifle can also be chambered for the .300in Winchester Magnum. *(Corbis)*

CHAPTER 4

FIELD CRAFT

Field craft is the skill of operating in a tactical manner that will reduce the likelihood of the soldier being killed and increase the opportunity for him to do damage to the enemy. It covers a wide range of subjects such as movement, formations, visual signalling etc. All soldiers learn field craft to varying degrees of sophistication, but for the sniper it is of paramount importance to his survivability and operational efficiency.

Camouflage

Camouflage is the part of field craft that relates to avoiding visual observation by the enemy through blending in with the surroundings. All infantry soldiers will wear a camouflage pattern uniform of some kind and this is sufficient to avoid the enemy's eye being drawn to him at a distance in the normal course of operations. This pattern will also tend to break up the shape of the soldier and make him a more difficult shooting target at range. He might supplement this uniform when fighting in a wilderness setting by attaching vegetation matter to his outfit, particularly the head area, as this is the part of the body that tends to be shown around cover or above ground.

The human eyes/brain combination is designed to pick up on familiar shapes. A rifle or a body lying on the floor will catch the eye much more if all of the object is the same colour, so the viewer can distinguish the entirety of the shape. The two main ways to break up a shape are to colour different areas differently, as achieved by wearing a camouflage suit, or to hide part of the shape from sight, keeping half of the body behind a bush, for instance. According to the sniper's operational situation, he might also wear a 'Ghillie Suit', an overall covered in suitably coloured rags and tassels which breaks up the soldier's outline into a diffuse pattern that easily blends into the landscape. The US Army *Sniper Training* manual describes how to make a modern version of the Ghillie Suit:

a. Ghillie suits can be made from BDUs or one-piece aviator-type uniforms. Turning the uniform inside out places the pockets inside the suit. This protects items in the pockets from damage caused by crawling on the ground. The front of the ghillie suit should be covered with canvas or some type of heavy cloth to reinforce it. The knees and elbows should be covered with two layers of canvas, and the seam of the crotch should be reinforced with heavy nylon thread since these areas are prone to wear out quicker.

b. The garnish or netting should cover the shoulders and reach down to the elbows on the sleeves. The garnish applied to the back of the suit should be long enough to cover the sides of the sniper when he is in the prone position. A bush hat is also covered with garnish or netting. The garnish should be long enough to break up the outline of the sniper's neck, but it should not be so long in front [as] to obscure his vision or hinder movement.

c. A veil can be made from a net or piece of cloth covered with garnish or netting. It covers the weapon and sniper's head when in a firing position. The veil can be sewn into the ghillie suit or carried separately. A ghillie suit does not make one invisible and is only a camouflage base. Natural vegetation should be added to help blend with the surroundings.

– US Army, FM 23-10, *Sniping Training*, 4-4 (1997)

A human face catches the eye too so, when a sniper is watching someone, it is a good idea for him to wear a camouflaged face-veil. This is a sort of string-mesh visor; from inside this hardly obstructs vision at all, but from the outside it totally hides the face shape and blends in with cover extremely well.

Snipers must be extremely disciplined to negate what is known as 'shine' – light bouncing off reflective surfaces, and thereby acting like a beacon identifying their position. There are very few items of shiny kit still in operational use amongst the professional armies. Brass buckles are a thing of the past, but polished boots, shiny black metal equipment and, worst of all, the soldier's sweaty face can still shine alarmingly when struck by light. A sniper should wear brown boots, paint his rifle in a disruptive pattern with colours suitable to the environment and make sure he applies cam-cream to his face. Even if he is really dark skinned, he should still use black or green camouflage cream to dull his skin.

Snipers should naturally be extremely careful when smoking, lighting cigarettes and reading maps by flashlight at night. It's a good idea for every soldier to be made to watch someone doing all of the above in darkness to see how good a target he makes. Like so many things, carelessness gets men killed. A sniper should not smoke at all on operations and should use a night-vision aid to read a map in the ambient light.

One particularly defining light signature of a sniper can be muzzle flash at night. Any soldier reading should take notice the next time he does a night-firing exercise, or goes on the range at night, how the muzzle flash lights up the whole area around the firing point. In a firefight at night, a sniper should move after each shot – wriggle backwards, roll sideways, then forwards to shoot again, or at least duck into a ditch or similar depression. After the sniper's shot, the enemy are likely to have their weapons trained on the point of muzzle flash, so the sniper needs to move away from this position. A sniper, of course, will only be taking one or two shots from the one position as a rule, but he should be aware that when he does take his shot at night he is telling everyone in the area where he is.

The effects of shadow are also of paramount concern to the sniper in the field. When in open ground and the sun is bright, a standing human will obviously cast a shadow. Furthermore, when the sun is above a person his shadow is small, but as the sun goes down in the sky the shadow it produces can become much bigger than the item casting it and, worst of all, it is black on the ground. Particularly when viewed from the air or from a hill looking down, the shadow is often the first and most obvious evidence of a person, vehicle or building, even if it is carefully 'cammed up' to be the same colour as its surroundings. A sniper should make sure there is higher cover right next to him so that he casts no visible shadow. When a sniper is viewing or shooting from a window, dark hollow or shaded area under a tree, he should make sure to sit well back from the sunlit area. At night he should also be careful to avoid placing himself in the direct glare of any artificial light sources, as these will also cast a long and highly visible shadow on the ground. At night soldiers manning defensive positions often use flares in an attempt to illuminate the ground around them. If a flare lights up, a sniper needs to hit the ground and freeze; wriggling around for a better position will get him noticed. As a general rule, the sniper must control any sudden movements.

At all times in the field, a sniper must not present his silhouette to the enemy. A sniper should look through cover, such as a bush, rather than around it, as this way he doesn't attract attention by presenting his outline. If he can't do this, then the best alternative is to look around the side of cover rather than over the top. When travelling cross-country, a sniper should also avoid walking along ridges. This option makes him visible in silhouette to anyone below him.

Smell as well as sight can betray a sniper's position. Particularly in the jungle or heavy bush, the sniper may sometimes be almost at touching distance from the enemy, and a man smelling of tobacco or after-shave may give a warning to the enemy and either spoil the shot or get the team killed. Native trackers in some parts of the world have uncanny olfactory abilities and can easily follow a man by his smell if he uses distinctively scented personal hygiene products or eats aromatic foods.

A sniper on an operation has to exercise complete discipline over all sources of noise. The metallic clang of a mess tin or rifle bolt sliding shut is a definite sign of a soldier's presence and will be immediately recognized by other soldiers. Prior to being deployed into the field, therefore, snipers will tape down any objects that could make noise, and avoid placing metallic objects side by side in packs and pockets.

On route to a shooting position, a sniper team needs to observe the rules of good spacing. The team members need to keep at least 4.6m (5yd) between each man, so that they present a more expansive target should they be engaged, rather than a clustered target that is easy to destroy. They should also vary the spacing as much as possible, as uniform spacing is easier to spot visually than a more random pattern.

When a sniping team leaves a shooting position, an overnight camp or an ambush position, they must be sure to take with them all evidence of their occupancy. What 'sign' they leave behind will tell the enemy a surprising amount about them. Even a good tracker cannot be totally accurate numbering a group of men walking in each others' boot prints, but if a few cigarette butts or food wrappers are left at each sleeping or firing position, this will often give away team numbers. The items can also indicate when the team left the position and in which direction, all extremely useful information to enemy trackers. Furthermore, a sniper should never carry anything personal that can give away the address or phone number of a loved one, including a mobile phone.

For the sniper on operations, one supreme fact is that the enemy needs to see the sniper to shoot, hence the principles of concealment should always be followed. When a sniper is travelling, he should try to stick to tree-covered ground, stream beds, dead ground or similar terrain. When a sniper is in a firefight he should show as little of himself as he possibly can. He must not climb up on the trench rim to get a better shot. Even when everything has gone wrong and the enemy know he is there and are shooting at him the soldier should make himself as small a target as he possibly can and release accurate, aimed shots at the enemy.

In many ways, the practice of sniping is an extension of the skills used in hunting. Stalking, whether it be stags or humans, is the skill of approaching an alert prey without being detected. The best way to learn this is to stalk instructors or other students across open grassy ground. In sniper school it is usual to start at 1,000m (1,094yd) and attempt to get within 150m (164yd) of the target who will be sitting on a truck or tower with a spotting scope. To make things more interesting, the targets have assistants called 'walkers' who are free to walk about the area under the radio control of the target as if they were counter-sniper teams. When the stalker/ student sniper gets within 150m (164yd) of the target, he is allowed to take a shot with a blank cartridge. Then he has to move and take a second shot from a different position. To prove that he reached the required distance, he is generally obliged to record the words in suitably small print written on a card which is placed by the target. A major problem here is gas from the rifle muzzle kicking up dust or grass, so the student should watch out for that and avoid firing from a position where the muzzle blast will give his position away. And he should remember that live rounds kick up a lot more dust than blanks. This exercise sounds challenging, and it is tricky, but it can be done and done well with practice. When a sniper can stalk a pre-alerted target like this he will find stalking a real enemy mostly straightforward, because in the real world most targets are not constantly expecting snipers to creep up and kill them. They have other things to think about.

There are several specific types of job which will be given to the trained sniper. Some of these involve sitting, as if in ambush, either hidden or in an overt position, waiting for the enemy to appear. The classic role of the sniper, however, involves making his way over hundreds of metres or tens of kilometres to get within range of a specific target before he takes his shot. To reach a target over any great distance, the sniper must be able to navigate with great skill, as not only must he find his way to the target but he must do so without being observed. This means picking and following a route which takes advantage of dead ground and thick cover wherever possible. For instance, a sniper might be given the job of killing an enemy general as he addresses his men at a specific grid reference. (A six-figure grid reference defines a specific box on a map which is 10m/32ft across.) Before he can take his shot, the sniper has to be able to navigate to that grid reference and cross the intervening country without being seen. In the navigation training provided to snipers by all armies, map reading and compass work is honed to a fine pitch with particular emphasis on the use of a compass. The reason for this emphasis on 'old-fashioned' equipment is that modern navigational tools, such as the GPS devices, can break down or be electronically jammed by the enemy, and the sniper must always be able to complete his mission and return to base. All sniper schools then go on to teach their snipers to work with a GPS, which is far quicker, more accurate and less prone to error than compass navigation.

A Sniper Mission

In order to present a clear example of good field craft and tactics in action, this section will look at the unfolding of a typical sniper mission. Although no two missions are the same, this section will provide a good platform to demonstrate the techniques snipers use on a daily basis.

Often the first stage of a sniper mission is target identification. If the sniper is tasked with looking for senior ranks as targets amongst the enemy in a conventional battlefield situation, then at the very least he needs to know the enemy rank structure and the various badges and stripes which denote these ranks, so that he doesn't shoot a corporal when he could be shooting a captain. More subtly, if he is working against insurgents in a foreign country, he needs to be able to recognise an enemy when he sees him. The Taliban in Afghanistan, for instance, tend to wear their turbans in a certain jaunty way that all the locals can spot easily, but which is invisible to a lot of Western soldiers. If a sniper has to look for a certain person or group of people,then he needs both photos of the target and the ability to memorize them. There is a game or exercise called 'Kim's Game', named after the character who played it in Rudyard Kipling's 1901 novel *Kim*. Kim was a boy in India learning to be a spy and play the 'Great Game', as the political intrigue between Britain and the other European powers was called then, and the game involved being given a brief period to memorize as many random objects placed on a tray as possible. The more remembered, the better the description and the more accurate the placing, the more points were earned. This game is still played in some special forces sniper schools, including the US Army Rangers, as it is good for strengthening memory for faces and for spotting when something in the field of view has changed.

Before the sniper actually sets out on the operation, he receives a briefing where the officer commanding the unit gives the sniper team their mission orders. These orders might be to sit in a hide on a roof top at a position in their forward operating base (FOB) and watch for an insurgent planting an IED, or it might be to walk 30km (18.6 miles) to a certain hill overlooking a road and shoot anyone that walks along it who looks like the enemy. A sniper might have to eliminate the highest-value targets he can see from a point given to him, or found for himself, within a certain area. This may sound vague, but the job of sniper can be very wide ranging.

During the briefing the sniper team will likely be warned of enemy forces known to be in the area and the presence and locations of friendly forces, if any. Friendly forces may also have been warned of the sniper mission to the extent of being told to keep out of a certain area, as is often done when special forces operations are executed. The officer giving the briefing needs to cover the following topics at the very least:

- The exact goal of the mission.
- The length of time the team is expected to be staying in position.
- The tactical situation where the team is operating.
- The time at which the team are expected to be at the final fire position (FFP).
- Emergency evacuation procedures in case of injury, illness or discovery.
- Radio procedures – frequencies, callsigns and codes.
- Friendly troops in area, where they are expected and what they know of the mission.
- The timing for exiting the FFP.
- Any special equipment needed for the mission, such as shovels or sandbags.
- Support, if any, in the sense of a patrol taking them to the FFP.

Once they know what and where their target is, the sniper team can begin to plan their mission. If they have been given the general area from which they will be shooting, they can begin identifying likely positions for the objective rally point (ORP), where they will regroup close to the target (see below), and for the FFP using contour maps, aerial photography, information gained from any units operating in the area and possibly the team's own visual reconnaissance before the mission. If the team has been given a specific patrol area, then it makes sense to work out 'way points' as stops along the route they intend to take around that area. Contours on the map can be used to select high points within the target area, which will give the team a good view of the roads or villages where the targets may be located.

Then they can plan to stay at each one for a little while in the hope of a target presenting itself.

At all times, the sniper team should be aware that they cannot rely entirely on the night or camouflage hiding their movements, as the enemy may have night-vision equipment or thermal imaging gear. The only sure way to avoid being seen by the enemy is to approach in dead ground, so the team should strive to put solid earth or rock between them and the enemy.

When a sniper team is going to be in an FFP for a long while, it makes sense to organize for themselves an ORP as a point of coordination. This ORP position should be well out of sight and earshot of the enemy and, if possible, defensible. The ORP serves both as a position from which to scout out and establish FFPs, and as a fall-back point for the entire team in case of an emergency.

The sniper team must remember that when they are on a mission there may be anti-sniper teams or enemy patrols about. This means they don't just have to worry about the target seeing them, but they must also keep a watch out for hints of enemy presence, such as areas of tall grass that move suspiciously, which could be an indicator of an enemy sniper approaching.

In terms of FFP selection, the position must give the sniper a good view of the enemy while providing him with as much cover and concealment as possible. The team should also be able to move into and out of the position without being seen, by selecting judicious approach routes. An FFP should also offer a wide field of view to give the sniper team as many target opportunities as possible. In some cases there may be a choice of targets, and the team may wish to take them all in quick succession; such is easier if the team do not have to move their firing position at all between shots. It is also a good idea to avoid selecting an FFP closer than 300m (328yd) to the enemy, as being in such close proximity increases the chances of detection following a shot. What is really a bonus is if the sniper can arrange for there to be some sort of obstacle – like a cliff or river – between himself and the enemy. This makes direct pursuit by the enemy immediately after the shot much less of a problem. The team should try to select a position where they are in the shade, and therefore in the dark relative to the observer, in the same way that a house interior looks dark through a window from the outside. Where the team is obliged to take up a position on a roof top or similar, they should try to put something higher behind them, even if it is only a piece of canvas, to stop their being silhouetted or to throw them into shade.

If at all possible a sniper should avoid choosing an FFP which is 'obvious' or isolated. He should never shoot from the traditional church steeple, for instance, as that is the first place the enemy will look for him, and they may just call in an artillery strike against the prominent feature. An anonymous place along a hedgerow, a random rock on a hillside or a random window in a block of apartments is far better. In dense cover there must not be twigs or grass between the firing position and the enemy, as when a bullet hits a blade of grass early in its flight the tiny deflection thus caused makes a great deal of difference to where the bullet travels over a long range. The most accurate shots are taken with the rifle resting on its bipod or on a solid support. To achieve this means lying down or kneeling with a higher support for the rifle in front of you. A position should be chosen with this in mind and every effort the sniper to avoid a situation where the sniper is sitting up and supporting the rifle himself if the range is anything more than 200m (219yd).

When the team get into the FFP, they should conduct a careful visual search across their entire field of view to spot any targets and enemy they may not be expecting. There may, for instance, be an enemy counter-sniper team sitting in clear view having a smoke. Once in the position the team will need to ask questions about what can be done to improve it. Does the enemy have patrols out? If so it may be best to build cover around the FFP from all directions. Does the enemy have air cover or reconnaissance drones? If so the team must have some kind of overhead screening. If the team are fighting a hi-tech enemy with infrared or thermal-imaging equipment, they will have to use sheets of the appropriate masking material (such as special blankets and clothing) in and around their position to hide their heat or visual signatures. In locations where the ground is very open, such as desert or light bush, the team might choose to dig in to reduce their visibility and provide cover from fire. When the growth of vegetation around the position is heavy, on the other hand, they might need to cut away any branches that are obstructing their view. In all cases, the team should have considered the country and their possible requirements in the planning stage, so they will have brought their entrenching tool or bow saw as required. Finally, the team should arrange their equipment so that it is easily accessible and set up a regime for watching, eating, resting and using the latrine.

As an aid to instruction, the types of FFP are listed in order by the time they take to construct. A 'hasty position' is used when the sniper team are only in a position for a short time or cannot reasonably construct a better position due to the enemy's movement or proximity. For example, the sniper team might bump into the opposition en route to the expected target, or might wish to set up an impromptu ambush. By definition, the hasty position has the following features:

- The sniper team uses basically what cover and concealment is available in situ.
- The position can be used for taking a shot in a very short time by removing small amounts of vegetation or backing a few feet away from the vegetation to conceal the weapon's muzzle blast.
- The position is not specifically selected or modified to offer protection from direct or indirect fire.
- The position relies heavily on the user's personal camouflage.

Occupation time with a hasty position should be no longer than eight hours.

When a team need to remain in position for a longer time than is safe or comfortable with the hasty position, an 'expedient position' should be created. This position aims to lower the sniper's silhouette as near to ground level as possible, but still allow him to observe and fire effectively. The expedient position is defined as follows:

- The position requires moderate construction, because it is formed by digging a hole in the ground just large enough for the team and their equipment. Soil removed from the position can be used to fill sandbags and these used in turn for building firing platforms or a firing rest, which needs to be solid – these can be covered by camouflage to reduce visibility.
- The position conceals most of the body and equipment.
- The position provides some protection from direct fire due to giving the team a lower silhouette and the cover of sandbags.
- The position normally affords little freedom of movement, but

more than in the hasty position. Team members can lower their heads below ground level, but slowly to ensure they do not produce a target indicator.

- The position allows little protection from indirect fire as this position has no roof to protect the team from shrapnel and debris falling into the position.
- The position exposes the head, weapons, and optics. The team must rely heavily on the camouflaging of these exposed areas.

Construction time is roughly 1–3 hours depending on the situation and occupation time is aimed to be 6–12 hours.

The 'belly hide' is similar to the expedient position but with overhead cover that protects the team by improved fire protection and also allows a little more freedom of movement. The position can be dug out under a tree, a rock or any available object that provides overhead protection and a concealed entrance and exit. The belly hide is defined by the following:

- The darkened area inside this position allows the team to move freely, as they cannot be seen. They must remember to cover the rear doorway so outside light passing through the hide does not silhouette the team inside and give the position away.
- This position provides protection from both direct and indirect fire. The team should try to choose a position that has an existing object which will provide good overhead protection or they will have to build it in the same manner as overhead cover for other infantry positions. This is generally a framework and sandbags.
- This position requires significant extra construction time relative to the above positions.
- This position requires extra materials and tools. The construction of overhead cover will require saws, waterproof material and so on. These will have been considered in the planning stage and will be carried by the team instructed to build the hide.
- The sniper team will have to lie in the belly hide without moving a great deal due to the limited space and design of the position.

Construction time should be around 4–6 hours and occupation time roughly 12–48 hours.

The 'semi-permanent hide' firing position is used mostly in defensive situations and requires equipment and additional personnel to construct. It is not an overt shooting position like a sangar; rather, it is another type of covert position. Yet it does allow sniper teams to remain in place for extended periods or until they are relieved by other snipers. The semi-permanent hide is defined as follows:

- The team members can move about freely within. They can stand, sit or even lie down.
- The position protects against direct and indirect fire.
- Loopholes are the only part of the position that can be detected. These allow for the smallest exposure possible by the shooter, yet still allow the team to view the target area. The loopholes should taper down to a small diameter on the outside of the position. Loopholes that are not being used should be covered from the inside with a piece of canvas or similar material.
- This position may be considered to increase the risk of detection, purely because having teams relieve each other in a position always increases this risk.
- This position is easily manned for extended periods.
- This position requires extensive work, personnel and tools to construct. It should be constructed during darkness and be completed before dawn.

Construction time might be 4–6 hours by four personnel and occupation time is expected to be 48 hours plus per shift or until relieved by other teams.

Whatever position the sniper team is in, the routines remain essentially the same. An acceptable rota for resting, eating and latrine calls must be arranged between the team. Latrine visits should be made during darkness as far as possible and human waste buried in the ground or deposited into a plastic bag and taken with the sniper, to avoid providing enemy search teams with a clear example of human 'sign'. Both sniper and observer should have a good firing platform whenever possible. The reason for this is that the sniper needs a stable platform for his weapon and the observer needs an equally solid platform for his optics. It is normal to rotate observation duties on a long mission, but the rifle should stay where it is and, where space allows, have the sniper move over to the spotter position to access the optics, log, range card and radio.

The team's actions after taking the shot depend to a great extent on the tactical situation and the mission type. If they are in an overt permanent hide then certain factors prevail. The enemy will know it is there, but not if the position is occupied. Once the shot is taken then they know the position is occupied for sure. If the sniper has just shot at a lone insurgent planting an IED then that individual ought to present no problem but are the team sure there is no counter-sniper waiting for a shot? If there is even a possibility of this threat, the sniper and spotter should get behind cover after a shot is taken. (It is better to be safe than sorry, given the price of being wrong.)

After taking the shot, the team should move from the FFP to the ORP as quickly as possibly owing to the potential response by the enemy. Even if the enemy has no hi-tech gear to pinpoint the team's position, they may well loose off some machine-gun fire, or a few mortar rounds, where they think the team might be. If the enemy do have professional surveillance kit, there will be accurate incoming fire of some kind certainly within minutes and possibly within seconds, so evacuating the FFP after shooting is an urgent priority.

The nature of a sniper's field craft changes significantly in an urban setting. A sniper team's survival during an urban conflict will depend to a great extent on the sniper's ability to place accurate fire on the enemy, while exposing as little as possible of himself to enemy observation or fire. To achieve this the sniper must constantly seek new firing positions after each couple of shots, and use them properly. He must be neither lazy nor careless in doing this. Positions available to the sniper in urban terrain are quite different to those provided by a rural area.

Urban terrain suits sniping very well and gives the sniper a marked advantage over other infantry. If they utilize the cover effectively, a sniper team, or group of teams with interlocking arcs of fire, have the capability to stop a fairly sizeable enemy seeking to advance through their area of responsibility. On the other hand, they must remember that the enemy will be very much aware of the sniper threat and will use every asset he has to detect and kill them. Over the years snipers have developed a number of principles concerning the effective use of available urban firing positions. For example, a sniper must be able to fire his rifle ambidextrously, to avoid having to show most of his body when the cover is on the wrong side for him. It is far safer to move back a yard from the brickwork in line with the target so as not to show the rifle muzzle around the corner.

Windows are obvious firing apertures for snipers, but they need to be handled with care, as the enemy will be scanning windows carefully. When a sniper fires from a window he must not allow the end of his rifle to protrude or expose himself to sight in the window frame. Instead he should stand well back inside a room where it appears darker, and reduce the angle from which he may be observed and can receive fire. Alternatively, he could create a loophole in a wall, although again he must not let the rifle muzzle protrude from the hole. The peak of a roof provides another excellent vantage point for a sniper, but it also silhouettes him against the skyline, so he needs to position himself amidst and in front of roof top objects to break up his visibility.

In an urban setting, the sniper should select his next firing position before he leaves the old one, to minimize the time he must spend in the open and moving. He should never fire from consecutive windows in a building, and must move after two or three shots, even if they missed, as detection is almost certain. Where more than one sniper team works in combination, they should operate from entirely separate positions, as two snipers shooting from close together are easier for the enemy to locate. Sniper teams in the same area should provide mutual support. This is particularly important when protecting each other from assault by infantry, with each team providing potential fire support for the other.

An escape route should always be planned and located before the sniper team enters an urban position. They should also be aware of potential hazards. For example, the enemy might seek to burn them out of their firing position, and therefore the team need to think of escape routes and other practical measures to reduce the fire risk. Garbage cans, crumbling walls, barking dogs and other urban hazards can cause injury to, or detection of, the unwary sniper team. Physical comfort, however, should be the sniper's lowest priority when selecting a firing position. Better to be uncomfortable for an hour or two than dead forever. Consideration should be given to selecting a position to which the enemy cannot easily gain access. A large building or terraced development may allow the enemy to enter at a distance undetected and approach the sniper team through the building and out of sight or hearing. Far better is a detached building to which an enemy approach must be over open ground. A sniper team should also attempt to get the greatest arc of observation possible to allow for the detection of targets and approaching enemy, while weighing this ideal against all the other factors, such as visual exposure to the enemy. The higher a firing position is within a building the worse the view it provides of the ground close by, and the more a sniper must show himself, say by leaning out of a window, to defend the building. Furthermore, the higher the firing position the slower the team can escape when the time comes. Aiming downhill from great height also affects the flight of the bullet significantly. For these reasons, the second floor is usually the best choice.

Another point in urban sniping is that a sniper should never attempt to fire at a distant target through glass situated immediately in front of the firing position, as this deflects the bullet a great deal. Where a target is close behind a distant glass window, however, the effect on the final impact point is insignificant. A sniper should open the window at his position, remove the glass entirely or replace it with plastic.

Controlling an Area

Snipers are becoming more and more popular with commanders in counter-insurgency situations, because their precision fire reduces the incidence of civilian casualties. While an artillery spotter *could* sit in a sangar and call in artillery fire against an insurgent mortar team at 2,500m (2,734yd), the artillery barrage would demolish all the houses and kill bystanders for a hundred metres around the mortar. A sniper, however, can just eliminate the individual members of the mortar team one by one. The same sensitivity to 'collateral damage' applies to the use of a sniper in place of a heavy machine gun, which is far more likely to kill bystanders too.

One of the most common and important jobs for a sniper in a counter-insurgency situation is controlling an area of ground. What this amounts to is watching an area to ensure no insurgent or enemy soldier can plant an IED, move weapons, move between positions, set up an ambush or take any other offensive or preparatory action. There are two ways of achieving this aim: covert control and overt control.

Covert control is where the sniper team is hidden and watching for an unsuspecting insurgent target to show himself. The insurgent does not know, initially, that there are snipers in the area. The shooting position might be in a town, but it is most likely to be out in the countryside watching a road, a village, a market or some other place where an insurgent might make an appearance. A target is generally recognized by acting suspiciously or very obviously carrying weapons or planting explosives. The team might even be watching the house of a known insurgent and waiting for him to do something clearly actionable, such as accepting a load of mortar bombs. This is pretty much the classic sniper/observer mission today and its performance ties together much of what has been covered so far. And, of course, besides shooting the insurgents the sniper team can call in an air strike, if enough enemy are together in one place.

The purpose of this type of mission is to take the enemy by surprise and kill them. Furthermore, once the enemy know a sniper is out there, but in an undefined location, the threat alone means the insurgent must either be more careful and slow in his movements, or refrain from operating in that particular area altogether. To make his presence felt, therefore, a sniper does not need to be there much of the time, just fire a shot occasionally so the insurgent knows he might be around. Ironically, this technique for dominating an area is very much a mirror image of what the insurgent does to the occupying force, which is to force extra security measures that slow operational tempo.

Overt control is applied when the insurgents know that a sniper is in a certain position. Very often the team is sitting in a sangar, or fortified box with viewing slits, on top of an accommodation block, watching the area around a unit's FOB, or they are in a specially built tower overlooking a major road junction. A sniper emplaced at a suitable position overlooking a road can prevent an ambush being launched, or an IED being planted, over a distance of up to 3.2–4.8km (2–3 miles). With a 360-degree view he is controlling an area potentially some 4.8km (3 miles) across. So a sniper in an overt position protects that position from enemy fire, makes operations difficult for the insurgents in the area he controls, and reduces the active insurgents in the area by elimination.

When a sniper is tasked with controlling an area, his job is normally to make sure no enemy operate within his line of sight and within a given arc of fire. This area of responsibility may be 360 degrees around a unit's camp or it may be a restricted arc of fire in one direction, such as an approach to a base or along a stretch of road. Normally the sniper on an overt mission will not have a choice of position. The upside to this mission, from the point of view of the sniper team, is that they will be sitting in a relatively comfortable outpost with a roof to keep the sun off and plenty of tea, food, smokes and company. The downside is that they are clearly visible to everyone around and the position will attract enemy fire.

The fact that the enemy know a sniper is sitting in the sangar guarding the base and waiting to shoot them has two effects on their behaviour. The first is that they will likely not make an obvious target of themselves close to the sniper. But to a non-sniper 2,500m (2,734yd) is a very long way, and out towards the edge of the sniper's range there may well be insurgents failing to take care and keep out of his sights. An insurgent decides to set up a mortar and drop a shell or two into a base and he forgets that 2km (3.2 miles) away a sniper is sitting in a sangar, watching him down his scope. So a sniper will get some 'easy' kills out towards the edge of his area, despite not being hidden. The second effect of knowing the sniper's position is that the insurgents might well try to kill him. Effectively they are then acting as counter-snipers, and the sniper team must exercise constant vigilance.

Looking at recent operations in Iraq and Afghanistan, there are three main types of weapon in the insurgent arsenal which are commonly turned against a sniper position. These are the Dragunov sniper rifle, the RPG-7 rocket launcher and any of the Soviet heavy 12.7mm or 14.5mm machine guns. The Dragunov is the most popular of insurgent sniping rifles (see the Appendix for more technical details). It fires the powerful 7.62 x 54mm round out to a designed range of 1,300m (1,422yd) and is semi-automatic. Generally it comes with the adequate PSO-1 Soviet scope. In truth, it is more a very good designated marksman's rifle than a precision sniper rifle, but at short to moderate ranges it is absolutely deadly. It has a higher rate of fire than classic bolt-action sniper rifles and if the insurgent sniper can get reasonably close he can bring heavy and accurate fire to bear onto the position. A double layer of sandbags will stop the bullet coming into a sangar and a screen hanging behind the defending sniper will stop his head or upper body being silhouetted and reduce his visibility. Yet the Dragunov's special armour-piercing rounds will go through body armour and any helmet a man can wear. The sniper team should therefore take care to increase physical protection where possible, and keep moving about when in the open to make a difficult target for the Dragunov-equipped insurgent.

The RPG-7 rocket launcher is another favourite of insurgents everywhere, with a range of 1,100m (1,203yd). There are actually a wide variety of RPG rockets for use against different targets, but insurgents tend to use the basic armour-piercing type. This shaped-charge projectile creates a thin armour-cutting jet of molten metal on impact, but projected into a confined room or space the blast will kill everyone inside. The warhead detonates on impact of the nose cone with anything solid, and the blast can penetrate more than 25cm (10in) of steel armour. Hitting a garage-sized sangar is no problem for anyone with one eye in working order, so the sniper's best defence against the RPG is the spaced-armour approach, forcing the charge to detonate at a distance which renders it ineffective. Such a defence is formed from two parallel and separated blast walls or an outer screen of wire mesh and an inner blast wall (the latter usually consists of metal armour, concrete or a sandbag wall). The way this defence works is that the shaped charge in the rocket is detonated early by the outer screen, the strength of the cutting jet dissipating before it reaches the main blast protection. Alternatively, the round is made to tumble and not focus its blast properly.

Both the 12.7mm and 14.5mm machine guns make mincemeat of sandbags and concrete walls at any range. Their weakness is that they are big, heavy and take some moving around. There is little physical defence against such weapon, so the sniper should be wary if one is known to be in the area. If such a gun does open up – it can be identified by the low and slow rate of fire – the team should hit the floor and crawl to safety as best they can. Or the sniper could try a shot against the machine-gun team, depending on the situation. Thankfully, these machine guns are valuable to the enemy and therefore have not often been used against sniper emplacements, because once fired they attract the typically destructive return fire of coalition forces.

Something that makes counter-insurgency operations complicated for the occupiers is that it is difficult to tell who the enemy is. In conventional warfare the opposition wear a different uniform and everyone can see which side they are on, but in a counter-insurgency operation the enemy rarely wear uniform and indeed take some trouble not to be identified. This is a problem for a sniper who is only observing people, often from a distance, as opposed to identifying them by drawing their fire.

The situation is made more complicated by virtue of the fact that in certain parts of the world the locals carry guns and knives. Many men in the Arab world carry ceremonial daggers in the front of their belts. All across the Middle East, Asia Minor and some parts of Africa, men and boys carry rifles, including the AK-47s of exactly the type beloved of insurgents everywhere. This means that just carrying an AK-47 does not necessarily mark someone out as an insurgent. What changes the situation is if an individual is carrying a weapon that is not culturally normal for self defence, such as an RPG-7, an RPK machine gun, parts of a heavier machine gun or parts of a mortar. Care needs to be taken with identification, however. In some places the locals grind corn with a round block that can look very much like a Russian anti-tank mine.

One obvious way of separating an insurgent from a neutral civilian is by his actions; an individual dressed in civilian clothes but emplacing an IED will naturally be classified as a combatant. In different theatres, the actions that identify an insurgent may vary slightly depending on their tactics. In Afghanistan, for instance, insurgents use men and boys as 'dickers' to spot troop movements and potential targets for them. Dickers very often stand out because they remain in one place for no obvious reason; this is a warning that either the insurgents are planning something or that they are present nearby.

A good example of a sniper mission in an insurgency environment comes from the following US Department of Defense (DoD) news report, issued in 2010:

FORT POLK, La., April 23, 2010 – It was April 2007 and the early-morning sky was clear as Army sniper Sgt. 1st Class Brandon McGuire and his spotter scanned for insurgents near Forward Operating Base Iskandaryia in Iraq. 'We were observing a stretch of road that had recently been cleared of IEDs [improvised explosive devices],' said McGuire, now the first sergeant of Alpha Troop, 1st Battalion (Airborne), 509th Infantry Regiment that's based here. 'The area had also seen a lot of mortar activity,' McGuire recalled. 'We had several soldiers killed and severely wounded along that stretch of road by IEDs and mortars.'

As the two soldiers gazed across the expanse between their hiding place – an abandoned shed – and their target area, they noticed a man in local garb sauntering up and down through a series of canals. When he dug into a canal bank and uncovered a mortar tube, McGuire knew this was a target.

'We called the battalion tactical operations center and reported what we had,' said McGuire, who hails from Olathe, Kan. 'We were granted permission to engage the target.' However, engaging the target was easier said than done. 'We measured the distance at 1,310 meters,' McGuire said. 'There was a crosswind of 8–10 knots and a sand storm was heading our way. We didn't have a lot of time.'

Yet time was needed. Snipers are trained to make a triangle from a target's chin to his chest, and then aim for that mark. But McGuire's target was moving up and down through canals, making it difficult for the Army marksman to get a clear shot. And, it was necessary to make calculations for windage.

But, McGuire caught a break – there were some children flying kites not far from the target. 'We were able to use the kites to help estimate the wind speed at the target,' McGuire said. 'We watched for almost two hours before the target presented himself in such a way that I was able to get a clear shot.'

McGuire said he didn't think he would hit his target with the first shot.

'I was hoping I would get close enough to make an adjustment and hit him with the second shot,' McGuire said. 'I knew that when I fired there would be a brown out for a couple of seconds – the dust would block my vision – so I was depending on my spotter to let me know where the first round hit.'

McGuire took a breath and then squeezed the trigger of his Barrett .50in caliber sniper weapon system. After the dust cleared, McGuire prepared for a second shot, but was unable to find the target. 'I asked my spotter, "Where is he?"' McGuire recalled. The spotter replied: 'I think you got him.'

McGuire said that for a couple of seconds there was disbelief on his part. Then it was back to work scanning the sector for targets. The shot was so effective that no one knew about it, other than McGuire, his spotter – and the target. 'Even the kids flying the kites were oblivious to what happened,' McGuire said. 'They just kept flying their kites.'

McGuire had removed an insurgent who had helped kill and wound American soldiers.

'No one knew who shot him,' McGuire said. 'Not even the local elders.'

The U.S. soldiers in the area gained an immediate benefit with the death of the insurgent, McGuire said. 'We'd had so many soldiers killed and who had lost legs,' he said. 'After the shot the daily mortar attacks and IEDs ceased in that area.' McGuire said 'the million- dollar shot', as it became known among members of his unit, was a big deal to coalition forces in the area. 'Everyone was congratulating me,' he said. 'But to me, it seemed like another day in Iraq.'

McGuire recently returned from a trip to California to film a segment of an upcoming History Channel special entitled 'Sniper: The Deadliest Mission', The two-hour documentary is scheduled to air this fall. 'I spent a couple of hours in an interview, then shot the rest of the day with another sniper,' McGuire said. 'It was a lot of fun.'

McGuire attributes the success of the improbable shot to tactical patience. 'It took us two hours to get the shot picture I needed on the target because of the terrain,' he said. 'We waited, then finally got the shot. Hitting a human target is not like a deer or something. With an animal, you can kind of predict what their movements are going to be, but with a human, you don't know what they are going to do.'

– Chuck Cannon, 'Army Sniper Recounts Amazing Shot', *Fort Polk Guardian* (23 April 2010)

This account is a fine example of the patience and intelligence required to take an effective sniper shot, and to make an accurate target identification in a complex insurgency conflict. Note that two hours elapses between spotting the target and taking the shot, a challenge to any sniper's powers of self-control. That the sniper finally chose the right moment is indicated not only by the instant kill of his target at long range, but also the fact that the civilian world around the kill was left relatively undisturbed.

CHAPTER 5
SNIPER TACTICS AND SCENARIOS

Although snipers can, and frequently do, operate in solitude on the battlefield, they are also tactically integrated with larger infantry units, typically a squad, section, platoon or company. This situation requires some historical context. Over the last 200 years, armies around the world have reduced, in a general sense, the size of a tactical unit of infantry used as a fire group, from a company of 150 men down to a section of around eight men. This is, in part, because the weapons issued to infantry have improved from muskets to automatic rifles with far higher rates of fire. Consequently tactics have changed from volley fire in lines, with each man capable of around 4rpm, to fire and movement in small groups, in which each soldier has a potential cyclical rate of fire of about 650rpm.

Today, the modern infantry section, of four to eight men, is required to patrol and operate alone with only radio contact to base and to fire-support assets such as artillery and air cover. Furthermore, since World War II infantry units in most of the world's armies have given up their accurate, long-range, bolt-action rifles and been equipped with some type of assault rifle. The Soviets opted for the reliable AK-47 7.62 x 39mm series and the Western nations were led by the United States who, involved in the close country of Vietnam, went with a high velocity 5.56mm-calibre weapon. Many European countries opted for the semi-automatic 7.62mm G3 or the FN FAL as well. The reasoning behind the adoption of the assault rifle has been to reduce the weight the soldier is carrying, in both weapon and ammunition, and increase the rate of fire. The latter advantage is at the cost of range, because in the close country in which future wars were expected to be fought the extended range was not so important, whereas the weight was an issue for every soldier. Indeed it was discovered during World War II that few infantry battles took place at ranges greater than 200–300m (219–328yd).

The Russians were the first to come up with the idea of an assault rifle in 1916, when they produced the Fedorov 6.5mm automatic rifle, designed by Vladimir Grigoryevich Fedorov. This weapon turned away from the idea of bolt action and maximum range to employ the principles of the assault rifle – a light weapon firing light ammunition from a large magazine, and with a high rate of fire but a relatively short range. The Fedorov saw service at the end of World War I, during the Russian Revolution and during the Winter War with Finland in 1939–40.

Then the idea of the assault rifle seemed to lie dormant for a while. During World War II, however, German engineers developed the first true assault rifle, based on research showing that most combat took place at ranges of less than 300m (328yd). The rifle they produced was the Sturmgewehr 44 (StG 44); in German *Sturmgewehr* literally means 'storm weapon'. Not only did it fulfil all the requirements of an excellent assault rifle, but it was specifically designed to be quick and easy to manufacture. It looked very much like the future AK-47, and saw combat service in 1944–45, but it arrived too late, and in too small numbers, to change the course of the war.

In Russia, one Mikhail Kalashnikov began his career as a weapons designer while he lay in hospital after being shot in the shoulder during the battle of Bryansk in 1941. From 1942 Kalashnikov worked for the Chief Artillery Directorate of the Red Army, focusing on small-arms design. He heard there was a competition being held by the military for a new rifle that would chamber the 7.62 x 41mm cartridge developed by Elisarov & Semin in 1943. One specific feature required of the new rifle was extreme reliability in the muddy, wet and cold conditions of the Soviet frontline, and the other stipulated specifications amounted to a call for an assault rifle. Kalashnikov, as the reader might surmise, produced the Kalashnikov AK-47 which looks and operates very much like the German StG 44. This weapon, chambered in 7.62 x 39mm, won the competition and went on to become the standard-issue rifle for the Soviet Army by 1949. It then went on to equip all communist bloc forces and numerous other armies and insurgent groups, becoming the most produced weapon in history. Still in use in its millions today, the AK-47 has had very few significant modifications since its inception. Taking everything into account, it is arguably the best rifle that has ever been made. It is cheap to manufacture, extremely reliable, is light, has a high rate of fire and the heavy bullet hits hard even at relatively long range. Certainly its performance is better than the 5.56mm weapons issued in the West. Though the AK-47 is referred to by soldiers in the West as the 'AK', it is known universally amongst Russian soldiers as the 'Kalash'.

The assault rifle was developed as the perfect weapon for 'intermediate' ranges, but at any range above 300m (328yd) the assault rifle has rapidly diminishing performance. For this reason infantry sections are not just equipped with assault rifles. With the advance in manufacturing technology allowing the mass production of excellent, relatively lightweight, machine guns from World War II onwards, it became the policy of pretty much all major armies to place a medium machine gun (MMG) of some 7.62mm calibre in each infantry section. By the 1990s, almost all infantry sections in the Western world were equipped with assault rifles of 5.56mm and an MMG of 7.62mm plus, of course, the usual anti-tank rocket launchers, 60mm light mortars, under-slung grenade launchers and so forth. Originally the purpose of the MMG was toincrease the firepower of the section, but a second advantage was that, as short-range assault rifles were issued, the machine gun increased the range at which an enemy could be effectively engaged, with the ability to engage targets usefully up to around 1,000m (1,094yd). The infantry of most armies are now equipped with assault rifles, so they are dependent on their machine gun for engaging any target out of their range. But, of course, there is only one weapon in each section which can shoot this far. The latest thinking, and that employed very often by Western armies deployed to Afghanistan for instance, is to replace the MMG with a light machine gun (LMG) firing 5.56mm rounds. This policy has two advantages: the first is that the weapon is much lighter for the gunner to carry, and the second is that the ammunition is interchangeable with the section rifle ammunition. There is one serious disadvantage, however, which has become noticeable only in recent years on the plains of Afghanistan. The LMG of 5.56mm calibre has a much-reduced range compared to the old 7.62mm MMG, and this reduces the distance at which the section can attack or defend itself. Though it is often claimed an LMG has a range up to 1,000m (1,094yd), at that range the rounds are well spread, losing hitting power and deflected by wind. So an infantry section on the ground, by themselves in wide open spaces, have an accurate defensive fire range of 200m (219yd) from their rifles and a little more from their one LMG. Of course, on a mounted patrol this thinking does not apply, as the infantry section will still most likely carry MMGs and automatic grenade launchers with a 1,600m (1,750yd) range, but they cannot carry these when they are on foot.

Any reduction in effective range is a serious problem when infantry are operating in an open landscape like that of Afghanistan. Though many Afghan insurgents have the AK-47 as their main weapon, this rifle, with its heavier bullet, still has the range advantage over the 5.56mm Western assault rifles. In combat Western soldiers are being engaged with some accuracy at more than 300m (328yd) and forced to advance towards the enemy, giving up the advantage of cover and risking mines, in order to return effective fire. Being slightly outranged in the open by AK-47s is bad enough, but a growing number of Taliban are carrying old bolt-action rifles such as the British Lee-Enfield .303in, and these have an effective accurate range approaching 1,000m (1,094yd). Worse, there are a growing number of Russian Dragunovs appearing in the field, which have a designed range of 1,300m (1,422yd), not to mention, of course, the fact that an RPG rocket has a range of 1,100m (1,203yd). So Western soldiers are either being picked off from a range at which they cannot effectively reply or they are being forced to take the weaker hand of assaulting troops to close with Taliban firing from prepared positions.

A controversial solution to this situation would be to scrap the 5.56mm round, as it is too light and short ranged. It should be replaced with a heavier round of 7.62mm or similar, with the consequent greater range, and this should be fired from a semi-automatic rifle which is made as light as a weapon can be to fire such a round. The superb German engineers at Heckler & Koch already make an excellent weapon called the HK 417, which would be an ideal replacement. The United States has the excellent 7.62mm M14 rifle, which served it well between 1959 and 1970. It is sufficient to point out that US military leaders are right now looking at going back to the 7.62mm-calibre rifle, or probably something just close enough to prevent the appearance of a U-turn.

The special forces of both the United States and the United Kingdom already use 7.62mm rifles themselves for infantry-type work. As a possibly interim alternative, it might seem that a sniper added to every infantry section is the ideal short-term answer to the lack of range in the modern infantry section. But actually it is not, as the sniper is already in short supply owing to the high skill levels he possesses. And, of course, the sniper has no useful rate of fire for close combat. Besides this, he has passed a long and very difficult course equipping him to stalk targets, build hides and use all manner of electronics, so as to fire his heavy, slow, delicate rifle very accurately out to distant ranges. A sniper is not at all what is required, even if sufficient were available. What would really fit the bill would be to issue all the section with a semi-automatic 7.62mm rifle, plus a marksman in each section with a more accurate semi-automatic rifle that uses the same ammunition as the other members of the section. The marksman would then be able to hit specific targets out to more than 1,000m (1,094yd). This would give the whole section a defensive circle of more than 300m (328yd) of accurate fire and 1,000m (1,094yd) of suppressive fire. Using this strategy the section could not be out-ranged by the AK-47 or the Dragunov or the RPG-7, and there would not be one slow-firing rifle in the section to weaken it. Plus all the weapons would continue to use the same ammunition for supply and exchange purposes.

Practical action is already being taken in the field. Some sections in Afghanistan are now being strengthened by adding a man with a rifle firing the same ammunition as the section, but accurized and with better sights. This improved rifle increases one man's effective range sufficiently to reduce, though not eliminate, the range problem. He is called the designated marksman.

The Designated Marksman

We have touched on the role of the designated marksman at several points in this book. Yet his role needs to be explained in detail to gain a full understanding of the sniper resources available to a modern army. In essence, the true sniper goes out into the field as part of a two-man team to hunt the enemy, or he guards a static position and prevents enemy activity in the surrounding area. He also works with intelligence operators on clandestine special operations. The task of the designated marksman is somewhat different. He is equipped with a rifle capable of significantly greater range and accuracy than his comrades' assault rifles. This rifle may be a standard-issue assault rifle with better sights, or it may be a specialist semi-automatic rifle firing the same ammunition more accurately. It could even be an entirely different rifle firing heavier, longer-range ammunition. His task though, whatever his rifle, is to work as an infantryman within a section of men, principally on foot patrol but also when mounted in vehicles. When the section find themselves engaged by the enemy at distances beyond the capability of their assault rifles, his mission is to engage and eliminate the enemy at ranges up to around 1,000m (1,094yd).

Importantly, the presence of a designated marksman should not reduce the strength of the patrol significantly as, unlike a true sniper with his heavy rifle and slow rate of fire, the marksman can still operate as an infantryman and produce effective targeted or suppressing fire at short range when required. Plus he can do sentry duty and dig a hole too. A true sniper needs to be a good shot, but he also needs a brain to deal with all the maths and technicalities of modern equipment. A designated marksman, however, just needs to be a solid infantryman and a very good shot.

A designated marksman's significance lies in his weapon, which has greater capabilities than the short-range assault rifles carried by the rest of the section. As we have seen, such was the thinking behind Dragunov's Snaiperskaya Vintovka Dragunova (SVD) sniper rifle, which is accurate up to about 1,300m (1,422yd) using the standard issue PSO-1 scope. This is plenty for the designated marksman role and, most significantly, sufficient for most sniper roles when they are operating just behind the frontline and picking off high-value targets.

It is also totally reliable wet or dry, clean or dirty. So the only fault, if it is a fault, is the lack of extreme range whereas the fact that it also doubles as a designated marksman weapon makes manufacture and supply far more efficient.

In recent years, confronted by tactical realities in Iraq and Afghanistan, Western generals have finally grasped the idea of having a designated marksman. The US approached the issue slightly differently to the Europeans. They continued to produce dedicated sniping rifles and worked on the problem of adapting or creating the ideal designated marksman rifle as a completely different tool to the sniper rifle. This seems reasonable, as the marksman's weapon is then ideal for its task as opposed to a compromise.

The US Army and the Marine Corps operate as totally independent units, with their own ideas about equipment and tactics. Yet as far as designated marksman weapons are concerned, both organizations have adopted, at least to some extent, an accurized version of the standard M16, known as the Squad Advanced Marksman Rifle (SAM-R) by the Marines and the Squad Designated Marksman Rifle (SDM-R) by the Army. The weapon fires the standard 5.56 x 45mm NATO round, but the rifle itself has been worked over and fitted with a scope to give accuracy out to a claimed 550m (601yd). The sharing of ammunition with the squad is a practical advantage, and it is also a useful to have communal spare parts and so forth. The M16 is a pretty good rifle, but the range is still short for the designated marksman role. Certainly if someone with a Dragunov is firing at a patrol from a hillside 1,000m (1,094yd) away, then the engagement could still become *very* one-sided in favour of the enemy.

It should be noted that the decision to use an accurized 5.56mm rifle was taken in a world where the infantry were issued with 5.56mm rifles, and so the people developing the designated marksman concept had to work within that remit. The Marines alone are now beginning to issue the M39 Enhanced Marksman Rifle (EMR), which is a semi-automatic rifle firing the 7.62mm NATO round and is accurate to a claimed 780m (853yd). That is going the right way. They are also beginning to adopt the US Navy's Mark 14 Enhanced Battle Rifle (EBR). This rifle was originally developed for the US Navy SEALs and is also a semi-automatic rifle firing the 7.62mm NATO round. Note that the EBR came into service in

2004 as a general combat rifle for the SEALs, a fact that serves to confirm what has been said above about the 5.56mm round being too light for open-country warfare. With iron sights the EBR is supposed to have a range of 500m (547yd), which may be optimistic, but with a scope it is probably good for the claimed 800m (875yd) or perhaps a little more. The logical next step would be to issue all infantry with the EBR and have an accurized version with even greater range. That would increase the range of the whole squad, allow the sharing of ammunition and increase the range of the designated marksman.

Following calls from soldiers on the frontline in Afghanistan for a rifle that could actually reach a distant enemy, the British Army also held a competition for a designated marksman weapon. This was won by a gas-operated, semi-automatic rifle of 7.62mm x 51mm calibre with a 20-round magazine. (Ironically this specification is largely the same as that of the old FN FAL/SLR rifle issued by the British Army to all infantry from the late 1950s to the mid-1980s.) This new rifle has four Picatinny rails with a 540mm (21in) top rail for night vision, thermal and image-intensifying optics, besides quick change barrels, various stock configurations and so on. It should be a good for a first-round hit at 1,000m (1,094yd) and sounds ideal for the task. All that is required to equip the infantry efficiently will then be a reversion to a 7.62mm rifle for all ranks.

This new British rifle has been called the 'Sharpshooter', but the official designation is L129A1. The Sharpshooter design, manufactured by Law Enforcement International in the USA, won a competition against Heckler & Koch's HK 417 (which is already supplied to specialist units such as the SAS), FN Herstal's Special Operations Forces Combat Assault Rifle (SCAR) and an entry from Sabre Defence Industries. Until the Sharpshooter comes into service with the whole of the British Army, the British soldier on the ground trying to fill the designated marksman role will have to manage with the old-bolt action sniper rifle, the L96, or use an SA80A2 with improved sights. The L96 is heavy and has a slow rate of fire, while the SA80A2 has not got the range required. As a poor and temporary fix, the L86A2 Light Support Weapon (LSW) has been marked as suitable for the designated marksman role because of its alleged increased range of up to 1,000m (1,094yd), alongside its capability of giving accurate automatic fire. In this situation, it would be expected to replace the Minimi light machine gun. The Royal Marines, the SAS and all other United Kingdom Special Forces, of course, already use the excellent 7.62mm HK 417.

Much of the time, in the counter-insurgency role, the designated marksman will be supporting the rest of the section as a rifleman. Occasionally he will be tasked by his section commander to target enemy soldiers who are out of range of the other members of his section. A situation may therefore develop where the designated marksman is required to operate more like a conventional sniper of World War II or the Vietnam era, and detach from the section or FOB to work more in the true sniper role.

Hostage-Rescue

Regular army snipers will tend to spend their operational life in conventional warfare scenarios, tackling either insurgent or general military targets. Special forces snipers, however, are required to respond to a far broader series of tactical challenges, including hostage-rescue missions. As we shall see, the hostage-rescue sniper rarely takes a shot at long range. The penalties for a missed shot are typically the deaths of hostages and bystanders.

Most domestic hostage situations resulting from burglaries or opportunistic/organized kidnappings are handled by SWAT-style armed police. In most countries, however, high-risk political and terrorist hostage situations have been dealt with by special forces for many years as evidenced by the Iranian Embassy siege in London, brought to an end by the SAS in 1980.

Terrorist groups take hostages for a number of reasons – raising funds through receiving ransoms, influencing government policy, coercing the release of political prisoners etc. Taking a hostage can be an effective action – various Palestinian groups have managed to exchange one Israeli soldier for a whole raft of their own men. Sometimes terrorist hostage situations are resolved by the police, but very often they are dealt with firmly by special forces using, amongst other tools, snipers.

In recent years special forces soldiers have also been called upon to deal with the threat posed by modern pirates. Pirates capture merchant ships and either steal the ship or hold it, the cargo and the crew for ransom. Piracy is a huge and often violent enterprise, with immense amounts of money being paid and huge vessels being seized. The main areas that pirates operate in are off the Horn of Africa and in the Far Eastern seas such as the South China Sea. There are a number of commercial security firms that protect merchant ships, not to mention the navies of many countries, but when a ship is seized one or more special forces snipers are often involved in the release of vessel and crew.

Hostage-Release Standard Operating Procedures

Terrorist hostage situations break down into two types. One situation is where a hostage has been taken by terrorists, spirited away and demands received by the family, government or media. The other situation is where the terrorists actually take control of a building and the people in it. These situations have significant differences when it comes to formulating an effective response, as might be imagined.

A defining feature of the former situation is that terrorists are trying to keep their hostage holding location secret. The most pressing problem for the authorities, therefore, is not so much releasing the hostages by force as finding them in the first place. For the terrorists, there is no need to keep the hostages anywhere in the locality or even the country. Very often they are taken away to a safe area controlled by the terrorists before any demands are made or before they are even missed. In countries where this sort of thing happens, from Israel to Northern Ireland some years ago, the occupying powers generally have some sort of intelligence operation in place and spies amongst the terrorist organization. If the hostages can be found, then the release attempt is usually a straightforward military operation against a building.

Observation positions are first set up around the hostage holding point and these are very often manned by special forces snipers. Then electronic audio-visual devices are brought into play, where possible, to listen to what the terrorists are saying, work out how many terrorists there are and possibly locate the hostages within the building. (The types of device used are classified, as that information might be useful to the enemy.) Next an assault plan is devised with the goal of storming the building to kill the terrorists before they have a chance to kill the hostages or detonate any explosives they may have. Special forces snipers may be used to initiate the assault or to eliminate terrorist lookouts to allow the main assault team to approach. If the ground allows several snipers to see targets, then they will be coordinated by the commander to fire together and eliminate the maximum number of terrorists in a single moment at the beginning of the assault.

It is a standard operating procedure (SOP) to designate two snipers per potential target, unless operational constraints make this impossible, as this allocation allows for casualties or a miss. The snipers will normally be equipped with silenced, sub-sonic rifles to avoid warning the terrorists that the assault has begun, the weapons having an effective range of up to 400m (437yd). Telephone lines and electronic communications will be jammed before the assault begins, but only at the last second to avoid warning the terrorists that the assault is in progress. Where terrorists remain alive after the snipers have opened fire, then the assault team will storm the building with grenades and short, handy weapons such as the MP5 submachine gun. Where possible, such an assault begins from the roof and works its way down through the floors to give the attackers the advantage on stairwells – it is easier and faster to make a downward-moving assault than to climb the stairs, and there is a reduced risk of grenades falling back on the assault team. Stun grenades may be used to disorientate the terrorists prior to each room entry, but rifle grenades fired from outside are more effective and less risky for the special forces soldiers. A grenade launched from a rifle-mounted, under-slung grenade launcher coming in through a window does not tell a terrorist there is someone outside the door of the room he is actually in, but he will still be stunned a moment later when the assault team does come through the door.

During the assault, the snipers will remain on watch to take targets of opportunity such as terrorists appearing in windows or attempting to escape the scene. When the building is secured, any survivors amongst the occupants will be handled with caution – sometimes terrorists try to pass themselves off as hostages, and each person will be positively identified before being released.

The vast majority of hostages taken by terrorists, or sold to them by criminals, are released following the payment of a huge fee or the release of terrorist prisoners. This is principally because it is so difficult to find the captives. In September 2007 two heroic Italian intelligence operatives, who had been working undercover near Farah, western Afghanistan, were captured by the Taliban. The Taliban's intention was to interrogate them to death and ransom their bodies back to the Italian government. The Italian military, however, were able to keep them under surveillance until a rescue mission could be organized. The two operatives were initially held in a safe house, but after four days they were moved in a convoy of vehicles, when the rescue was mounted by the British Special Boat Service (SBS).

From what we know, the British team, from C Squadron, arrived on the scene in four Lynx Mk 7 helicopters and snipers immediately opened fire from the airborne helicopters with .50-cal rifles, first to disable the vehicles by shooting out their engines and then to disable the terrorists themselves. As this was happening, two of the helicopters landed and the soldiers surrounded and approached the vehicles, killing the remaining insurgents as they because visible. It was a classic example of special forces' proficiency, and also demonstrated an innovative use of snipers.

Sometimes a terrorist group will take over a large building such as a school or an office block, or a vehicle such as an aircraft or bus, and hold it with the people inside as hostages. They will then attempt to defend the building or vehicle against assault by the military of that country. The rules for this type of operation are different to the previous situation because the authorities have a clear sense of the terrorist location, and the terrorists are aware of that fact. The terrorists also know that they are almost certain to die in the event of a military assault, so they will be intending to take their hostages with them. Sometimes the hostage-rescue forces can take back an objective with minimal civilian cost, such as in the Iranian Embassy in London where the SAS killed all but one of the six terrorists for the cost of two hostages (one was killed prior to the assault). Greater tragedy occurred at a school in Beslan, North Ossetia-Alania, in 2004, when the Chechen Muslims held 1,100 people hostage, mostly schoolchildren, and killed over 380 of them before they were overcome by Russian security forces.

The first stage in any overt hostage-rescue mission is for the authorities to throw a cordon around the site. This will prevent bystanders getting in the way, stop support reaching the terrorists and also prevent their escape. If the site is on board a grounded aircraft, then it is already somewhat secure, being on a runway, but if it is a building in a city then care must be taken to isolate it. The first response units on site are likely to be the local police with an armed team. They will be given orders to secure the site and do nothing to provoke an initial confrontation. An exception to this rule is often found in Russia; the police may be allowed to shoot any terrorist who presents a target of opportunity. The Russian Spetsnaz, their special forces, also have a policy of storming any target site if the negotiators have not succeeded within 48 hours. The reason for this rapid intervention is that the Russian leadership are aware that the entire purpose of many such kidnappings is to generate publicity, so ending it quickly reduces that benefit and motivation.

The second thing that happens is that the negotiators arrive, probably a specialized team attached to the police. The negotiators' primary objectives are to obtain the release of some of the hostages, focus the terrorists on discussion rather than violence and control the debate between terrorists and authorities. The hostages are as much a burden on the terrorists as an advantage – the terrorists are fully aware that should they begin killing hostages, or the negotiators think they have run out of options, the shooting may start and their lives be cut short. And much more important than killing the hostages, from the terrorists' point of view, is to draw out the event as long as possible to stay in the news. So there is a likelihood that some hostages will be released, a few at a time and after long delays, to extend the time they are in the media spotlight. The negotiators' task is to get as many hostages released as quickly as possible before the shooting is started by either side.

The hostage-takers will have typically planned how they will capture and hold the premises. Certainly they will have firearms, but they are very likely to have explosives with them too. They will mount guards on all approaches to the site and in some cases mine or booby trap these approaches to obstruct rescue. The terrorists may also wear explosive vests, of the suicide-bomber type, and might even put these on the hostages. The whole site may be wired to explode on a command button held by the chief terrorist. It is very likely that, where circumstances permit, the hostages will be split into groups at different locations within the building in an effort to make their rescue more difficult.

Soon after they arrive at the hostage site, the negotiators will make contact with the terrorists and find out what they are asking for. The police commander and/or the chief negotiator will keep senior politicians informed about the situation. Major countries all have special forces response teams on 24/7 alert for this type of event and these will arrive within an hour or two. They will be briefed on the situation by the police, and then deploy snipers, intelligence personnel, technical specialists and the assault team. The terrorists will be expecting the authorities to jam communication from the site to the outside world. Telephone lines will be electronically 'cut' and electronics will be employed to jam radio and mobile telephone signals. Snipers will be placed around the site, initially as observers to gather intelligence, but also ready to shoot on command.

As the negotiators begin communication, electronic intelligence-gathering devices will be covertly installed around the vicinity of the building and, if possible, within the building itself. Then the waiting begins. The operational commander will have his team observe and negotiate with the terrorists. Their goal is to build up a picture of the terrorists' habits and organization while getting the maximum number of hostages released before the shooting starts. Each side is very likely to be aware of the other's objectives. While this is happening, the assault team will be on standby in a requisitioned building close to the scene and the snipers will stay in their positions, gathering intelligence and awaiting the order to shoot. The assault team may begin to rehearse the potential assault plan at this stage.

As the siege continues water, food and medical aid will be supplied to the terrorists and hostages, and small concessions made in an effort to obtain the release of hostages either by group (i.e. children) or in ones and twos as special cases. This is a joint effort between both sides, as the negotiators want as many hostages released as possible before the assault, and the terrorists are content to release some hostages to postpone an attack. By these means sometimes all the women, children, sick or old are released and sometimes they are not. The snipers will by now have given the commander a good idea of how many terrorists are available to be shot at any one time, and if any of these appear to be the leader. Electronic means may have ascertained the numbers and positions of terrorists and hostages within the building. This is all taken into consideration as the plan for hostage-rescue develops.

The commander will be well aware it is only a matter of time before the shooting starts. For political reasons, and to minimize the news coverage, he will usually be under pressure from his political boss to end the siege quickly. Certainly, once the terrorists begin killing hostages the order to launch the assault is often given. The snipers are immediately given the order to fire on their selected targets, killing visible terrorists, then continuing to watch for targets of opportunity. The assault team storm the building, entering through the roof and systematically clearing the building room-by-room. Anyone who runs from the building is arrested, searched for weapons then placed in an area away from danger for later identification. If all goes to plan, within a few minutes quiet falls within the building and the assault is over. The remaining hostages are led out of the building by the assault team. A search team enters the premises while the cordon remains in place, looking for further explosives and terrorists. Finally the hostages are debriefed as an aid to the identification and prosecution of any surviving terrorists, and in preparation for the next hostage-taking event.

Fighting the Pirates

Pirates are nothing more than maritime criminals out to make money and prepared to kill to do so. Yet they are reaching almost military levels of mission planning and execution. In certain parts of the world, these criminals have found that they can board commercial shipping and take the crew hostage before ordering the vessels to head for the pirates' home port, often a harbour in a failed state such as Somalia. Once safely in port, the pirate leaders contact the ship's owners and demand a huge ransom for the release of ship and crew. For a large ship, the ransom will be some millions of dollars. At the time of writing, there are more than 20 large ships and more than 400 crew held by pirates in Somalia alone.

The men who arrange the piracy are rich and organized criminals who control operations from their well-guarded villas secure from local government interference (there is no government to speak of in Somalia). The pirates who do the boarding of the vessels are poverty stricken fishermen who are involved to make a wage. Once a vessel has been taken to a receptive port, it is pretty much lost as the hostages are dispersed, the ship is wired to explode and it is well guarded.

Stopping piracy depends on the target ships defending themselves with firearms and a naval force mounting a rescue as soon as possible following a successful hijack. For some time the navies of a number of nations have patrolled the areas most subject to piracy, with the mission of preventing pirate attacks, but this policy has had only limited success. The problem here is that the sea is a big place, and it is impossible for patrolling warships to watch every square kilometre of sea. What they can sometimes do, however, is send helicopters to the rescue of ships which radio that they are under attack from pirates, providing they are close enough to the situation. Yet the only really effective solution is that each ship should carry an armed security detail able to defend itself until such times as a naval patrol sends support.

The seaborne pirates, as we have seen, are controlled by a head man on shore and it is he who finances and organizes the logistics of the piratical operation. In essence this consists of a 'mother ship', which might be a captured deep-sea fishing boat or similar, and a number of high-speed open boats called skiffs. The mother ship cruises offshore until the onboard pirates spot a suitable target. On occasions the target is spotted by lookouts on a different ship, whose crew then proceed to tip off the pirates.

When it has approached to within several kilometres off the target vessel, which is typically a large container ship or oil tanker, the mother ship launches 4–6 skiffs packed with armed pirates and these approach the target at speed. Where the target ship has armed protection aboard, the defenders show their weapons or fire a few rounds into the air and the pirates turn away. If there is no protection crew, or the target vessel is entirely unarmed, the pirates proceed to attempt a boarding of the ship.

Only very recently has the British government allowed British flagged vessels to carry armed protection and other nations vary as to their rules and policy. A friend of the author in command of an unarmed vessel protection party has had his men manufacture wooden replica machine guns and paint them black. This worked on one occasion, but it will not continue to be effective indefinitely, as the pirates are not fools and are beginning to force the protection party to fire rounds before they stand off. There is asteady trend towards carrying armed protection teams in troubled areas, but this is at the discretion of the shipping companies at present. As only a relatively few ships are hijacked out of the many thousands plying the seas, many owners are preferring to save the $50,000 dollars a week which the protection costs. At the moment a team, where one is aboard, usually comprises four armed men, three where the ship owners are seeking to cut corners.

There are many less obvious things which a merchant ship can do to protect itself from boarders before using lethal force, given the training and will. These include increasing speed, manoeuvring, deploying fire hoses and emplacing barbed-wire barricades. And when they all fail there is the final option of opening fire on the pirates. The reader will accept that when seeking to board a huge ship with a side like a cliff from open boats travelling at 15 knots (28km/h or 17mph), the advantage lies with armed defenders. Boarding an unarmed vessel is difficult but achievable given the fit young pirates employed for the job, but doing so against an armed vessel is virtually impossible if the guards are alert. But there is another tactic open to the pirates. If the ship's guards are armed with assault rifles, the pirates can stand off 1,000m (1,094yd) and shell the ship with RPG-7s until they are allowed aboard.

Now we have seen what the piracy threat involves, we can take a look at what is being done to defend the world's shipping, by using three theoretical but representative scenarios for how an attempted hijacking might take place. A commercial freighter has been shadowed for two days by an alleged fishing vessel, which the master fears is a pirate. The fishing vessel is lowering skiffs into the water at a range of 4.8km (3 miles). The master has called for help and a French frigate, at a distance of 193km (120 miles), has launched a helicopter to offer assistance. The helicopter arrives at the freighter's location before the pirates board to find five skiffs full of armed men approaching the merchant vessel. Once the skiffs sight the helicopter they turn around and head back to the mother ship. For all their determination, pirates will almost never provoke a fight with a formal military force, and the aerial surveillance will mean that the element of surprise is lost.

The helicopter must decide what to do, or at least the captain of the frigate must decide, as fuel considerations will mean that the time over the suspicious boats is limited to some minutes. They are not, of course, allowed to sink the skiffs or even fire warning shots unless the crews commit any crime. A sniper with a 12.7/14.5mm anti-material rifle, for example, could shoot from the helicopter to disable the skiffs' outboard motors, but this action would expose the captain to charges of piracy for attacking an 'innocent' vessel.

There is the danger now that once the helicopter has pulled away the pirates will return to the attack, safe in the knowledge that there will be no interdiction before they have secured control of the vessel – the helicopter will need to refuel before it can head out again on patrol. From this scenario, it is evident that unless a vessel is positively identified as being of piratical intent, there is little patrolling military vessels can do, apart from be a protective presence in the vicinity of vulnerable vessels.

So it is clear that there are limitations to what military forces can achieve in pirate-infested waters. The following scenario, therefore, looks at the realistic defensive options for an armed vessel. The year is 2013 and most commercial vessels have armed guards aboard. The pirates have begun targeting defended merchant vessels, but their tactics have changed. They now lie off the beam of vessels carrying inflammable cargo and fire RPG-7 rockets at them until the master allows them aboard. A large commercial tanker carrying liquid fuel has sighted a group of skiffs approaching, and the master has alerted the protection team but also radioed for help from any naval vessels in the vicinity. None are close enough to render assistance within two hours, by which time it will all be over one way or another. Fortunately, the onboard protection team have a pair of long-range sniping rifles amongst their equipment, giving them a defensive radius of at least 1,000m (1,093yd). They begin taking shots at the pirate skiffs even before the threat comes within

RPG range. In a gentle swell a large tanker is a very stable platform for snipers, and they are soon registering hits on the approaching vessels. Very soon the pirates turn away and depart; the cost to the protection team is nothing more than a couple of dozen rounds of target ammunition.

The war against the pirates is one of continual measure and counter-measure, and even protection teams cannot be a total guarantee against the threat of hijack. Yet they certainly tip the advantage in favour of the commercial vessel. The next scenario illustrates not only how limited protection can expose the vessel to pirate attack, but also how a strong military response – if delivered in a timely fashion – can switch the situation around.

The master of a German freighter is trying to please his employers by keeping down costs, and part of this economy drive involves running without armed protection. Despite the crew's use of water hoses and barbed-wire barricades, a group of pirates have gained access to the deck and bridge of the freighter and taken the ship under control. Some of the crew are required to sail the ship back to Somalia, remaining at their posts with drug-taking pirates watching them vigilantly, armed to the teeth. The remainder of the crew are corralled in the mess hall, aside from one man who was shot and wounded during the boarding – he has been thrown over the side. Fortunately the master was able to get away a message before the pirates came aboard, and a British frigate has despatched a Lynx helicopter to render assistance. The helicopter is carrying a detachment of 10 Royal Marine Commandos, together with an SBS sniper team, and they are keen to arrive. When the helicopter does reach the scene it takes up station hovering ahead of the vessel. A number of pirates are visible on deck and they fire AK-47s at the helicopter from an impossible range. The special forces sniper aboard the helicopter takes down the four visible pirates in quick succession, then the helicopter moves immediately to hover over the fore deck and drops an eight-man boarding party by rope. These secure the deck by fire and movement under cover of the sniper, then take possession of the bridge. The pirates choose to surrender at this stage in preference to being cleared from the accommodation of the ship by rifle and grenade.

The scenarios above are all based on actual tactical possibilities and events. A recent real-world anti-pirate operation further illustrates how pirates are no serious opposition to a well-trained special operations team. Captain Richard Phillips, 53, was in command of the freighter *Maersk Alabama* and its 19-man crew as it sailed off Somalia in April 2009. The ship carried no armed protection, so when pirates appeared off its bow they were able to board the ship without serious resistance.

As they boarded, Captain Phillips locked his crew in a safe- room and offered himself as hostage to the pirates. He managed to send off a radio message in order to raise the alarm. This was the first time a US citizen had been taken by hostages, and the US Navy wasted no time in sending the guided-missile destroyer USS *Bainbridge* to the rescue. As *Bainbridge* made its way at full steam for the *Alabama*, a team of SEALs were deployed aboard the destroyer. Yet by the time the *Bainbridge* arrived at the hijacked vessel, the pirates had transferred Phillips to a covered, motorized lifeboat in which he was guarded by four men. This led to a standoff that lasted five days, but eventually three of the pirates were in view at the same time, and one was holding an AK-47 to Phillips' head. The SEAL snipers were ordered to open fire and killed the three pirates with three simultaneous shots, each of which hit a pirate in the head. The fourth pirate was captured and Captain Phillips released unharmed.

Of course, this effective action raised fears for the 230 hostages held in some dozen ships captured by pirates and anchored off the coast of lawless Somalia at that time. Angry pirates vowed to retaliate: 'From now on, if we capture foreign ships and their respective countries try to attack us, we will kill them [the hostages]', Jamac Habeb, a 30-year-old pirate, told the Associated Press from one of Somalia's piracy hubs, Eyl. Another Somali pirate claimed this rescue would escalate the situation: 'Every country will be treated the way it treats us. In the future, America will be the one mourning and crying.' Abdullahi Lami, one of the pirates holding a Greek ship anchored in the Somali town of Gaan, also told the Associated Press, 'We will retaliate (for) the killings of our men.' On this basis, it appears that the anti-piracy sniper and assault teams will be kept busy for the foreseeable future.

Covert & VIP Operations

A special forces sniper will almost certainly be involved in operations around the world which his government chooses to keep quiet. There are many small wars being fought all the time and Western governments secretly offer help to some of the parties involved, to gain commercial or political concessions.

Before the launch of the 'War on Terror' in 2001, Western armies as a whole were not seeing a great deal of combat, and a man had to join the special forces of his nation to experience military action. That action consisted mostly of being lent out to help a small friendly nation deal with an insurgency situation. Even more secretive, however, has been the realm of 'Black Ops', military operations conducted in such as way as to be deniable by the commissioning government. Sometimes this intervention involves a Western government supplying arms unofficially to a developing country or 'freedom fighters' based there, without paperwork and through a middle-man arms dealer. Or it might be financing and supporting a coup attempt by 'mercenaries' in Africa or elsewhere. Sometimes it means Western special force soldiers pretending to be a mercenary force. There were, for instance, some familiar faces in North Africa recently when the Arab Spring uprising occurred.

Sometimes overseas direct action means either protecting a VIP or politician who is very unpopular in his native country, or even killing one who is unpopular with a Western government. These situations are not mere speculation – they happen a lot more than an outsider might think. To keep the following discussion accurate, therefore, only operations in which associates of the author have been directly involved will be discussed. First we will look at the challenge of assassinating people overseas, and then at the issues involved with protecting a person at risk of such assassination.

The planning and technique required for every assassination operation is different because of a whole range of factors: where the target is located, how well he is protected, the general state of the country itself, police and communications, population density and so forth. Scheduled airlines, the road network and the type of people who live there determine how difficult it is to get operators and equipment into place, for instance. So as far as there is such a thing, the following is the standard game plan for an assassination operation using a special forces sniper.

The decision to carry out the operation is first made by politicians, with military and intelligence advice, at a very senior level. The pros and cons of removing the VIP are weighed in the balance for their political, social and international effects. At this stage there is normally no decision made as to how the assassination will be brought about, merely whether it should or should not occur and if so whether it should be overt, covert or a 'foreign flag' operation where it is made to look as if a convenient third-party state or criminal made the kill. Shooting is only one method available to a government and has both advantages and disadvantages according to the situation.

Given a positive decision to act, the intelligence people then gather available information on the target's movements and habits and so forth, to aid them in making a decision on the area where the operation will be carried out. As a rule, they will have been doing this for some time already on the off-chance that a strike would be authorized.

When precise information about the target's movements and habits in a foreign country is required, drones (more officially known as unmanned aerial vehicles/UAVs) are limited in their usefulness so where possible a ground photographic reconnaissance team is sent to the area tasked with discovering as much about the target's personal life as possible. Photographs can then be used to guide the elimination team. Of course, the active government will already have electronic surveillance of the target's telephones, computer use and email. Many powerful states operate covert intelligence-gathering operations in both friendly and unfriendly countries and this is normally low-level intel controlled from the local embassy wherever possible. Where this is not possible because foreign nationals of the acting government would stand out or are not welcome, then intelligence handlers control local agents from a relatively secure position, such as over a friendly border. To avoid any misunderstanding, in all intelligence operations the active government's operators themselves are not normally the actual information gatherers; they are 'handlers' for agents recruited locally by bribery and/or blackmail, who can gain access to classified information owing to their family, job or position. Whatever means are available, the active nation will gather information about the target's movements, identification photos if possible, potential shooting sites and routes in and out from these locations.

Using the intelligence gathered in this way, a team of experienced military and intelligence officers will develop a plan or plan options to eliminate the target, taking into account his location habits, protection and so on. The shooting team, perhaps four or more men or women and including the sniper team itself, will be summoned to a briefing and following this they may be held incommunicado to prevent leaks. The shooting team will then be moved into the area by means suitable to the situation. Depending on the country, its development and its defences these means may be anything from chartered flights to a submarine or a microlight aircraft. The weapons and electronic equipment will usually be brought to the site through diplomatic channels when a chartered flight is used for entry.

The shooting team will establish a situation much like a conventional ambush, with the sniper team as the 'fire group', ready to eliminate the target, and perhaps two of the team acting as 'stops' – sentries watching for the approach of the target. One member will usually be on communications with a scrambled satellite phone to receive live data from HQ, such as warning of local forces, and perhaps video from a drone, depending on the situation. Care has to be taken with the use of the satellite phone, as not only can its location be plotted, but it can also be monitored unless the conversation is encoded. Encoding in itself will attract the suspicion of local intelligence operators, particularly if satellite phones are not common in the country.

When the target finally makes his appearance, the shooter takes his shot and, if the sniper's aim is true, the job will be done. Assuming that planning is done properly, the shooting should be a relatively simple task. The most difficult part of the operation is typically the extraction of the team after the shot is taken. Obviously, shooting someone does not look like an accident, though there may be some doubt as to which organization is responsible. For this reason, it is most sensible not to shoot at very close range to make the ensuing search more difficult. In most cases weapons and communications equipment will be disposed of immediately, followed, perhaps, by washing the shooter and courier in chemicals to remove weapon-related traces. Diplomatic immunity may then be used to get the team out of the country or they may be evacuated by other means.

Having worked through the principles of assassinating a VIP, we now turn to the sniper's role in protecting a 'principal', which is the name given to the client of a close protection detail. There are a great many companies around the Western world founded and staffed by retired special forces personnel for the purpose of providing armed security services in high-risk countries like Afghanistan. These firms are always interested in well-paid assignments and less concerned about the ethics of the mission than some people might be. For there are a whole range of potential attackers depending on the principal's identity and where in the world he or she is being guarded. These range from a crazy person in the crowd armed with a pistol, through to a well-organized strike by a foreign state or terrorist organization. Of course, much of the protection does not involve snipers. A sniper will not normally be charged with protecting a film star or banker, unless there is a tip off that a terrorist plot is in motion, but a special forces sniper may well find himself working as part of a team covering the state visit by some foreign politician to his country, or the visit of one of his country's politicians to a foreign country. There are always elite snipers on the roofs wherever the President of the United States makes an appearance and the same is true for many of the heads of European countries.

When a political dignitary or head of state makes a public appearance, there is a whole army of people who work on different aspects of the security. Some principals, like the President, take their whole security team with them when they travel. The author was asked to assess the security for a recent US Presidential visit to London and discovered that besides the Secret Service protection team and a whole platoon of special forces snipers, he had two armoured helicopters for travelling in, Apache gunships to overfly the crowd when he appeared and F-15 fighters to control the airspace above him. This military entourage is in addition to the huge numbers of professionals who gather intelligence, monitor electronic traffic, search under the roads wherever he will drive and much more. An experienced sniper is by far the quickest way to neutralize a physical threat in close proximity to a principal. Alternatives such as agents rushing to the scene might be far too slow to prevent an attacking sniper in a tall building or a pistol-armed assailant in the crowd from making an attack. The sniper's brief is simply to take down anyone he sees with a weapon threatening the principal, as soon as authority is given by his commander. He will always be in radio communications with a controller to give the green light; sometimes the sniper will see the target first and call it in to control, sometimes another operator who will pass word to the commander.

Besides a lone gunman or bomber, there is the possibility of a coordinated armed assault on a VIP, such as happened in Egypt on 6 October 1981 when President Muhammed Anwar al-Sadat was reviewing troops in a stand amongst lots of other senior military figures. Suddenly a group of soldiers who had been in the parade leapt out of their vehicles and charged the reviewing stand, firing automatic weapons and killing the President. Sadat led a Muslim country and was trying to make peace with the Israelis, and the attackers were Muslim fundamentalists opposed to this idea. Two sniper teams could have stopped this assassination with ease. Benazir Bhutto, twice Prime Minister of Pakistan, was assassinated at a political rally amongst a crowd of supporters on 27 December 2007 in Rawalpindi, Pakistan. The motive was to prevent her spreading democracy in her country. She was travelling in an armoured Range Rover, but stood with her head and shoulders out of the sunroof to wave to the crowds. The first assassin approached with a pistol and fired three shots at her, then a suicide bomber approached and detonated himself close to the vehicle. The cause of her death was a skull fracture caused by the concussion of the blast blowing her head against the vehicle. Snipers mounted on roof tops, out of range of assault rifle fire, would have been an ideal protection against this attack.

Counter-Sniper Tactics

In this final section on the skills required of special forces snipers, the focus shifts first to counter-sniper tactics, and then to how a special forces sniper avoids being targeted by a counter-sniper. (To avoid misunderstanding, 'counter-sniper tactics' refers to the drills employed by all soldiers to avoid becoming casualties of attacking snipers, whereas a 'counter-sniper' is a sniper tasked with shooting an attacking sniper team.)

The sniper must deal with the counter-sniper threat principally, and obviously, by avoiding detection. A special forces sniper may very often be at risk from enemy counter-snipers when he is on task, but he is also at risk from plain general snipers when he is back in base camp resting. In some situations, he may be more at risk than other soldiers owing to the high value the enemy will place on killing either any special forces operator or a sniper in particular, identified by insignia or weaponry. Defending against the attentions of enemy snipers falls, therefore, neatly into two parts: general precautions against snipers and avoiding enemy counter-sniper teams. We will look at each of these in turn, building on the discussion above concerning general security at an FFP.

Defenders can keep snipers out of range of their base or convoy using a number of tactics, including aggressive patrolling in vehicles and on foot, the deployment of friendly counter-sniper teams and the use of drones to provide aerial reconnaissance. The central idea is that the defenders should make stalking to within range a risky and perilous policy. The two occasions when this tactic does not work so well are when the nearby terrain provides plenty of cover for a sniper and in built-up areas where a shot can come from any window.

Soldiers on patrol should stay out of sight when they can, and when they can't hide they must keep moving. At extreme ranges camouflage does help to break up the sniper's view, so it is worth some trouble getting it right. On foot patrol in a built-up area the best a soldier can do is to adopt an erratic pattern of walking called 'hard-targeting' by the British and perfected in Northern Ireland. The idea is that by taking two paces forward, one back, one to the side and so on at random the soldier makes a long shot very difficult for a sniper

vehicle and can cleverly differentiate between incoming and outgoing fire. It can provide range and bearing to targets or present them on an LED display. It is relatively cheap, reliable and very useful at close range, so most useful in built-up areas.

There is a more advanced system using the Boomerang concept, one that operates through four microphones mounted on the helmet of each soldier in a squad. By clever computing this machine gets around the inaccuracies of GPS and uses the range and bearing from the known position of each soldier within sound of the shot to work out the sniper's position to within one degree of arc and a couple of metres range. The advanced Boomerang 3 System, which has been integrated with the Land Warrior equipment system, then superimposes the position of the sniper onto the lens of a monocle worn by a soldier. The reader will understand how difficult it must be to cause the apparatus to still point out the sniper when the soldier has dived for cover and thus moved his position.

Lastly, there is a system in development called Red Owl, which uses microphones plus lasers to determine the precise position of a sniper. This has already been mounted on the PackBot Robot, which is a tracked robot unit the size of a dog. PackBot is normally used for sniffing out explosives or deactivating IEDs. Set up with Red Owl, however, it follows infantry around and points out anyone shooting at them. This may be starting to sound like something from a science fiction movie, but it has been tested in combat and found to work reliably in many situations.

Once an infantry unit has identified the source of sniper fire, the unit needs to shoot back at him as quickly as possible, owing to the fact that he may be either readying for another shot or crawling away as quickly as he can. He certainly will be moving away if he thinks he has been spotted. The thoughtful sniper will always be in dead ground, of course, and unexposed to direct fire if he possibly can be. The best way to shoot back at snipers, therefore, is to use mortars or artillery if the unit is out in the countryside and collateral damage is not an issue. A barrage of shells all around the area the sniper fired from is likely to prove deadly to a man out of cover. By the time this work is published, radar-fused mortar bombs will be more available and the airburst option will get the sniper pretty much whatever he does. In towns, though, response to a sniper is more tricky. Artillery cannot realistically destroy a city block because there is a sniper in it, at least not officially. So what can be done? In the past infantry under attack from a sniper have tried to get him to make a second shot with a dummy head on a stick, and then have their own sharpshooters take him out when he showed himself. But today there is something far better and it is called the XM25.

The XM25 or Counter Defilade Target Engagement System is midway between a rifle and a grenade launcher, and fires an explosive 25mm shell with a range of 700m (767yd). But it is so much more than either of these weapons. It is held like a rifle and the special recoil dampening system allows it to fire a heavy projectile at quite a high velocity, which produces the fairly flat trajectory required for accuracy. The really revolutionary thing about the XM25, however, is that it has a laser range-finder built into the weapon which talks to the shell that is about to be fired. So if a soldier carrying an XM25, or his sniper detection kit, sees a sniper shoot from a window over the road, he can take aim at the window and the weapon automatically gets the range. Then the operator presses a button and adds on, say, 2m (2.2yd) to that range as the distance at which the shell will explode. He fires at the window, the shell then goes through the window, and a built in timer explodes the shell at exactly the right moment inside the room, destroying anyone inside – the shell has a 5m (16ft) kill radius. The principle is that the laser range-finder combined with a radio-controlled explosive shell allows the shooter to set the shell to explode by the enemy's head in mid air after passing through a window or penetrating a wall, or indeed when it is over a trench. As this is written at the end of 2011, the XM25 is being issued to some US troops. The sooner this excellent weapon is available to all Western troops the better.

The hi-tech counter-sniper equipment becoming available is currently changing the way snipers are fought in all theatres. This is not to say that all the old tactics of World War II are no longer used or relevant, but it does mean that many of the risky SOPs should soon be consigned to the history books. One such suicidal piece of counter-sniper advice is known as the 'Rush', also known as 'Close With And Destroy'. This dictates that if a squad is taking casualties while being pinned down by sniper fire, then the order may be given to rush the sniper's position. If the sniper is too far away for a rush at him, then a rush-to-cover may be performed as an alternative. The principle is that the squad is taking casualties anyway and will take more if they stay where they are, so as the sniper has only a slow rate of fire it is less costly in men to charge him, or for all to run together to a better position.

The pincer movement tactic, by contrast, is applied if a squad are a rural situation and know where the sniper is, but cannot hit him with direct fire, and they do not have indirect fire available. In the pincer movement, a squad or pair of squads can move through unseen, preferably via dead ground, to drive the sniper towards another squad. This may sound very much like a flanking attack without the benefit of covering fire to the reader.

The fact remains that some snipers are going to get through and make hits. Of course not all hits are instant kills. There are many catchy sayings designed to drive home the idea of how valuable the first few minutes are for saving a man's life when he is shot and seriously wounded. Phrases like the 'golden hour' and the 'diamond five minutes' spring to mind, but the truth is that it really is the care a casualty gets in the first few minutes after being wounded that determines if he is going to live or die. Treatments like stopping severe bleeding, clearing the airway and other simple medical procedures make all the difference to casualty survival. To this end infantry NCOs should make sure that everyone on their team is well practised at first aid, and then whoever is still on his feet when the sniper has done his worst can patch up the ones who are not. If an infantry unit is so fortunate as to get a real medic assigned to them, then they should treat him or her like gold dust.

A counter-sniper team, indeed any sniper team, have many disadvantages when facing the enemy. They have very low firepower and can be easily destroyed in a face-to-face contact with even a small unit at close range. They have little protection from the weather, and physical hardship limits the duration of a mission and can compromise the team's security as hide discipline becomes lax. But the team do have two important advantages, one of which is surprise – the enemy does not know they are there. Being concealed gives the element of surprise to the sniper team in the same way as a efficiently sprung ambush. It also gives the team the advantage of firing on an enemy accurately while the enemy cannot effectively reply. The other advantage of the sniper team is that they can fire at an infantry unit from beyond the range of that unit, as we have already considered in relation to insurgents armed with Dragunovs. When facing a sniper team, a counter-sniper shares the same disadvantage of rate-of-fire but all the advantages of range too.

An infantry unit will probably not call in, or be allocated, a specialist counter-sniper team unless they are taking heavy casualties and are unable to stop the sniper or snipers who are harassing them, even by using heavy indirect fire and the other means outlined previously. Unfortunately, this sort of situation is occurring quite regularly in Afghanistan at the time of writing. Western troops are dug in at well-defended FOBs, from which they head out on patrol. Taliban snipers are targeting them both in the FOB and when they go out on patrol, and the effect is sapping morale in some areas to an alarming extent. There is little worse for the spirit than a soldier seeing his mates killed or wounded around him one by one over a period of days or weeks. The standard first response to sniper harassment is to use the available detection systems and then bring down an artillery, mortar or air bombardment onto the place the sniper is shooting from. Sometimes this works and sometimes it does not. As the reader has seen, the skilled sniper has already moved after taking his shot by the time the bombardment begins.

There is no doubt, however, that a specialist counter-sniper team adds a significant security and intelligence capability to regular units. Such is made clear by this 2007 Department of Defense story from Afghanistan, which also explains something of the arduous sniper training plus the close working relationship between sniper and spotter:

KIRKUK REGIONAL AIR BASE, Iraq, Nov. 19, 2007 – When servicemembers here have to go 'outside the wire,' they sometimes have an extra set of eyes watching over them.

Concealed and sometimes from a long distance away, the members of the 506th Expeditionary Security Forces Squadron's Close Precision Engagement Team, also known as the Tiger Team, observe, provide intelligence and, if necessary, neutralize threats.

The Tiger Team consists of Air Force security forces counter-snipers whose expert marksmanship and ability to practically stay invisible allow them to sneak up to an enemy undetected.

British Army AW50 (Arctic Warfare) bolt-action .50in sniper rifle. Note the adjustable butt support for stability. *(Corbis)*

Special forces weathermen can be crucial for supplying information on battlefield atmospherics, so that snipers can accurately assess and compensate for the conditions they are firing in. Here, a US special forces soldier looks through a laser range-finder. *(Corbis)*

A key part of the training for all snipers is learning to use binoculars, as shown here on the ranges at the US Army Sniper School. Note that the high-power binoculars are all on stands. *(Corbis)*

A sniper team pictured here during high-angle mountain sniper training, during which shots must be taken without a weapon support. *(Superstock)*

ABOVE: This Advanced Combat Optical Gunsight has a red dot in the sight picture which continues to mark the bullet impact point whatever angle it is viewed from. The system is currently being issued to the US Marines because it is faster and more accurate than using iron battle sights in close quarter combat. *(USMC)*

BELOW: A sniper's view of his target through the crosshairs, Dubrovnik, Croatia. *(Corbis)*

Shooting positions are here demonstrated by sniper, and trainer of snipers, Colour Sergeant Michael Bruce McIntyre BEM of the British Parachute Regiment. *(Courtesy of Michael McIntyre via the author)*

An American sniper peers at a potential target through the scope of his M24 rifle while on an operation to ensure a prisoner exchange goes well. He is inside a roof space and looking out under the eaves of the dwelling. *(Corbis)*

A rebel sniper in Tripoli, Libya, readies his Dragunov during the anti-Gaddafi uprising in 2011. *(Getty Images)*

Members of the US Secret Service sniper unit (The Presidential Bodyguard) keep watch during President Bush's visit to Daytona International Speedway. *(Corbis)*

A British sniper from the Irish Guards gives cover during an operation to neutralize insurgents and put out oil well fires in Basra, Iraq, April 2003. *(Getty)*

A British sniper team from the 3rd Battalion, The Parachute Regiment, conduct a strike operation in Kandahar Province, Afghanistan, in 2008. *(Getty)*

A US Marine scout/sniper team spotter scans a ridge for insurgent activity during an operation in the foothills of the Hindu Kush, Kunar Province, Eastern Afghanistan. *(Corbis)*

A US Marine scout/sniper, in winter camouflage, ejects a spent cartridge case from his M40A3 bolt-action sniper rifle during mountain training. *(Corbis)*

A US Army Special Forces sniper wears a 'Ghillie Suit', including vegetation as camouflage, to hide while waiting for his target. This effort is ruined by failure to camouflage the straight, black rifle barrel. *(Corbis)*

Kandahar Province, Afghanistan, 2010. A US soldier uses his helmet to attract sniper fire. This patrol was sent out to neutralize a mortar, supported by a sniper, repeatedly directed against their FOB, attacks that caused 20 casualties amongst the 150 men in the unit. *(Corbis)*

As part of the hand-over of Iraq, a US soldier instructs an Iraqi soldier on the use of a Barrett M107 sniper rifle. Tal Afar, Iraq, May 2006. *(Topfoto)*

'A large part of our job here is reconnaissance for the Army and sometimes agents with the Air Force Office of Special Investigations detachment here,' said Staff Sgt. Curtis Huffman, CPET noncommissioned officer in charge.

'When they have a mission outside of the wire, we'll set up near that location about an hour or more before they get out there. Concealed and out of sight, we are able to observe the area and give them real-time intel before they even arrive,' Huffman said.

Through direct communication with the mission commander, the sharpshooters let the team know how many people there are in the area, their exact location, if there are any weapons, or if the people seem to be hiding anything. That way, they know exactly what to expect before arriving at the location.

'Close precision engagement provides us with the ability to see into the future,' said Special Agent Christopher Church, OSI Detachment 2410 commander. 'They provide us with a situational awareness that we would not have without them. Having them watch over us during missions makes an enormous difference.'

The sharpshooters' skills also help save lives during counter-improvised explosive device and counter-indirect fire operations. 'We respond to routes that get hit by IEDs a lot or an area that is known for launching IDF (attacks),' said Huffman, who is deployed from Eielson Air Force Base, Alaska. 'We'll set up somewhere concealed along that route or that area, where we can watch people setting stuff up so we can get them before they can hurt our guys. We could be there from 24 to 72 hours.'

CPE team members also respond to their own comrades. If security forces members on patrol or on a post perceive suspicious activities in the area, they can call on the team to come out and, using their trained eyes, optics and night-vision capability, determine if there is an actual threat.

Each sniper team consists of two people – the spotter and the shooter. The spotter's responsibility is to determine things like the distance to the target, wind direction and then provide the shooter with corrections, which are adjustments on the rifle.

'Spotters do all the mathematical equations for range estimation, windage, everything from start to end,' said Airman 1st Class Matt Leeper, CPET member, also deployed from Eielson. 'The spotter definitely has the more difficult job. Your spotter has to be quick and accurate when giving the corrections. There is no time for the shooter to think twice. Your spotter is always right.'

The Air Force has about 350 trained sharpshooters. To become a counter-sniper, one has to be a security forces member, have proven marksmanship abilities and accomplish three weeks of training at Camp Robinson, Ark. 'The school is physically and mentally very challenging,' Leeper said. 'You are learning from the first day you get there. The first few days are in a classroom, and then you are on the range shooting.'

This is where the students are introduced to the M-24 sniper rifle, the military version of a Remington 700. 'The trigger squeeze on this weapon is a lot lighter than the M-4, and it also has a lot more kick,' Leeper said. 'Your shoulder gets roughed up at school, where we fire more than 100 rounds a day.'

Though shooting is only a small part of their job at Kirkuk, it's often the most important aspect. 'Only about 5 percent of our job is taking that shot, and the other 95 percent is intelligence gathering,' he said. 'But when you are in a situation where you have to neutralize a threat, you can't really think about anything except you have positive ID on that target, they have a weapon or you know they are placing an IED. You put that target in your crosshairs, you imagine it's just a blank target at your school house and you pull the trigger. You don't have time to think about anything else.'

The counter-snipers accomplish many missions at Kirkuk, but they find the most rewarding thing is being able to watch over soldiers or OSI agents. 'This is the reason why I joined,' Leeper said. 'When we are out there giving them info and providing cover, I feel like I'm doing my job. I don't feel like I deserve a medal – nothing like that. This is what my job is and what I joined to do. I joined to come to Iraq, and I went through sniper school to be an asset to the Air Force.'

– Staff Sergeant Markus M. Maier, USAF, 'Air Force Sharpshooters Watch Over Troops in Iraq' American Forces Press Service (19 November 2007)

One particularly notable statement in this passage is that concerning the percentage of time the sniper team spend in intelligence gathering. In a frequently fast-paced operational world, the value of having individuals spend hours quietly observing terrain and people is great. At the same time, there is a definite protective element to the counter-sniper work, the muzzles of the sniper rifles providing a vigilant defence against the enemy shooter.

To reinforce the worth of a good counter-sniper team, there follows a counter-sniper scenario based closely upon operational realities in Afghanistan at present. There is an FOB deep in insurgent-controlled countryside and surrounded by lightly covered ground and the odd village. For the last three months a platoon of infantry have been holed up there and taking casualties from snipers and IEDs. The roads and tracks in the neighbourhood are mined and booby trapped, so patrolling is desperately dangerous. Several men have lost their legs to IEDs already.

Roughly every week, on average, a sniper takes a shot at the camp. He seems to be a skilled sniper, as two sentries are now dead from his efforts and a couple more men are seriously wounded. The FOB's sentries are now using periscopes to look out from their positions to avoid the sniper's attentions, but the men going out on foot patrol cannot. It is assumed to be one man doing the sniping owing to the level of skill he is showing. He does not hit the camp regularly – maybe twice in a few days then not again for a fortnight so the defenders never know when he is in the area. And each time he works from a different spot. After each sniper attack there is a counter strike with mortars, guided by the electronic sniper detection equipment, and a patrol goes out to check the firing position. He has been firing at a range of around 500m (547yd) from holes cut in the walls of buildings, from a bank by a stream under some trees, and the previous week the patrol found he had killed a snake with his knife while sitting in position. Clearly the sniper is highly trained, skilful and very cool. The defenders' morale is ebbing because they cannot see a way of stopping him.

The officer commanding (OC) the unit reports the situation to his battalion commander and a special forces counter-sniper team are sent to the FOB to deal with the problem. The SAS counter-sniper team supplied is actually four sniper teams under the command of a sergeant. The first thing the sergeant does on arrival is minutely examine the area and the places from where the sniper has fired already. This gives him an idea of how the sniper is thinking when selecting a position. As he is always shooting from a range of around 500m he has formed a pattern and given the team a way to catch him. The sergeant draws a line on the map around the camp, making a circle 500m out. This line passes close to all the positions the sniper has so far fired from. Next the sergeant marks likely sniper positions all around the clock face, centred on the FOB but further out from the camp than 500m, and sends out his teams to recce them. By the following day they return from an apparently ordinary clearing patrol, on which they were dressed and armed as the occupying troops, with a selection of potential sniper hides. Each of these hides is spread around the clock face and is suitable for a sniper team to lie undetected for an extended period, watching the enemy sniper's next potential choice of firing position. The sergeant examines these positions on the map and aerial photos before selecting four positions spread around the clock face.

Late the next night, four members of the counter-sniper team leave camp each leading an infantry patrol carrying digging equipment and sandbags. They move quietly out to the selected hide positions and prepare a hide for each team. This task takes much of the night, but they make a long patrol after the digging and return to base by different routes. The following evening, loaded like mules with their equipment and supplies for a long stay, the counter-sniper teams themselves leave camp after dark and accompanying infantry patrols. They break off along the route to take up position at their hides, then the patrols circle the area and return to base. The cautious deployment is because most FOBs are watched day and night by the enemy, and it is important they do not know that there are counter-sniper teams going on station, or that a patrol has gone out and not returned.

The days pass and the teams settle into a routine of watching and waiting. There are near misses in terms of discovery, but the hides have been carefully chosen to avoid the local goatherds stumbling upon them. Nothing happens for eight days. Then, on day nine the sniper takes his shot and a sentry at the FOB is hit in the shoulder. The shot is fired from under an abandoned and burned out car a hundred metres from counter-sniper team Charlie. Of course they cannot see the shooter but his shot, and therefore position, is picked up by the electronics. The infantry loose off some mortar bombs in that direction to avoid giving the impression there is any change in their tactics. For the counter-sniper teams it is hard to believe that anyone could have got to the nearby position without being seen. Yet clearly they have, so a team shooter adjusts his position to cover the vehicle and waits. After the third mortar bomb explodes on the hard ground, sending a cloud of dust into the air, a slender figure dressed like any other local, but cradling a Dragunov rifle, crawls quickly out from under the vehicle. It is not a difficult shot at this short range and the heavy-calibre bullet kills the sniper instantly.

This type of scenario has played itself out dozens of times in Iraq and Afghanistan over recent years, and in other conflicts ranged around the world. The special forces snipers are capable of widening the protective radius around everyone from regular soldiers to VIPs, while also taking the fight directly to a frequently elusive enemy.

CHAPTER 6
FUTURE SNIPER WEAPONS

Having looked thoroughly at the tactical considerations of the special forces sniper, we now turn to the future, and examine the weapons that will define the art of the sniper over the coming decades. The current state of sniper technology is established by the latest record shots being out at around the 2,500m (2,734yd) mark, achieved with a 'classic' bolt-action rifle. Clearly kills significantly beyond this range cannot be achieved with just a slightly better chemically powered projectile weapon, as this technology is already close to its limits and the effects of wind drift and air distortion are becoming the limiting factors. We do know, however, that there are some very heavy-calibre weapons becoming available, like the 14.5mm and 20mm machine-gun derivatives we have seen earlier, which are threatening to smash the current records and bring in perhaps a 5,000m (5,468yd) kill. But surely that must be about the limit of what can be done with a normal rifle? In this section the reader will see that this is not actually the case: the killing range of the conventional rifle is being stretched even further by the development of bullets which home in on the target. But to reach a great deal further than 5,000m (5,468yd), there needs to be a switch to a different type of technology. The choice seems to be between continuing with something 'fired' from the shoulder of a sniper, or opting for an extension of the 'calling in fire' concept where a sniper would use some sort of electronic aiming device to direct fire from artillery, air assets or even space craft onto the target.

The advance of technology does not stop with replacing just the weapons either. There is a move already towards replacing ground troops with robotic ground vehicles which can either be controlled, like the Reaper UAV with its bombs and missiles, from a shed somewhere totally safe, or even think for themselves as they attack the enemy. There is no good reason why similar remote-controlled or autonomous technology could not be adapted to replace the sniper with a machine on many missions. But the sniper does not have to fear for his job just yet. Though the use of a sniper robot has some advantages, such as endless patience and the removal of all risk to humans, it is not clear how such a robot can sneak into position across a rural landscape, or even into a roof-hide in a town, anywhere near as inconspicuously as a human. So probably the special forces sniper will keep at least the glamorous jobs for the foreseeable future.

Over the last 30 years or so, technology has moved forward at a rapid pace and many things have improved in all areas of operations, such as communications, fire control, air support, body armour and night vision. Over this same period the sniper's job has increased in importance dramatically, owing to his ability to avoid inflicting collateral damage on a civilian population during a counter-insurgency campaign, but his equipment has not improved to the same extent. Yes, his rifle is a little more accurate and his sights are a little better than years ago, but when an idea has been around for a long time it gets to a point where improvements slow down and the only way to move forward is to change tack and develop an entirely new path or idea.

An example of this principle is the motor car. From the early 1900s to the 1990s it developed at an amazing rate in terms of speed, range, comfort and reliability and then over the last couple of decades we have just seen slightly better fuel consumption and tweaks in body shape. In the opinion of this author, the conventional sniping rifle has reached a similar point to the motor car of the 1990s, and the time is fast approaching where there will be a major shift in the type of weapon issued to the sniper so over the following pages we will consider what form this shift might take.

We have already seen that there are natural limits to what can be achieved through conventional rifle design in terms of range and accuracy, owing to wind drift, light distortion and mechanical imperfections. The way to gain maximum range and accuracy, all other things being equal, is to use a heavier-calibre bullet fired at a higher muzzle velocity, as this round will be less affected by wind and imperfections of manufacture. Of course, there is a limit to the weight of rifle a man can carry and therefore, because rifle weight restricts bullet weight, a limit to the maximum size of bullet which he can fire at the enemy. And, whatever the calibre, above a certain velocity, a bullet encounters all manner of weird effects, such as breaking up in the barrel and generating heat through air friction and then melting, besides slowing down faster and delivering diminishing returns for the effort of pushing it so hard. A bullet travelling twice as fast, of course, generates four times the air resistance owing to the inverse square rule we touched on earlier. These physics-related factors mean the bullet fired from a rifle will not get a great deal bigger or travel a great deal faster than it does at present, so we have already reached the maximum practical accuracy inherent in the conventional rifle and bullet combination.

Rifle manufacture itself is not likely to improve a great deal either. This is because, with the development of modern engineering techniques, a quality rifle barrel is now made pretty much as accurate as it can be, and a competition or sniping round is made close to perfection too. Production has become such a precise art that by far the most important things that affect the accuracy of any specific rifle and bullet all happen after the bullet has left the barrel. Having said that, this currently achievable mechanical accuracy is very good. The best rifles made today have an MOA of about 0.2, which equates to 5cm (2in) at 1,000m (1,094yd) or so in calm air when fired from a bench clamp. Unfortunately, a stiff breeze can easily deflect such a round by tens of metres on its way to the target and that deflection will constantly vary according to the shift in the wind. The reader can see, therefore, that the major limiting factors in rifle accuracy today are either in the aiming of the rifle, causing it to make better allowance for wind and sight issues, or in dealing with the wind on the way to the target. The fact remains that the extreme man-hitting range for a *conventional* medium-calibre rifle will not rise much above 4,000m (4,374yd) and a heavy-calibre rifle might make 5,000m (5,468yd) or 6,000m (6,562yd) owing to wind effects or the distortion of the sight picture brought about by heat or other factors.

One way of getting around the effects of wind deflection, and light distortion, on the bullet as it travels towards the target is to guide the bullet in flight. In this way its course can be corrected and a hit might be achieved at any range the bullet can reach. The EXACTO programme is just such an initiative. It is a development programme set up by the US government to produce a .50-cal bullet which is capable of extreme accuracy at long distances against moving targets. The idea is that this programme will address the fundamental limitations to rifle accuracy which are, principally, variables in the environment such as wind effect. Here is a synopsis of the patent application relating to the winning entry for the EXACTO programme (it has been simplified a little):

> A small calibre (no more than .50in) laser-guided bullet which has a self-contained guidance system. This includes on-board laser sensors and navigational circuits which are capable of determining the deviation of the bullet from the optimum trajectory travelling along which it would impact the target. This technology will generate an electrical signal to control piezo electric steering surfaces on the bullet to effect a course change in the bullet. The guided bullet uses a number of symmetrically arranged laser sensor elements which are positioned about the longitudinal axis of the bullet. These sensors transmit optical radiation signals from the laser target beam to photo detector elements built into the bullet. The signals are then amplified by logic circuits to produce the functions necessary to operate the steering surfaces on the bullet and guide it onto the target.

As the bullet is fired, and while it is travelling, the shooter keeps his aim on the target and a laser beam is projected from the rifle to the target for the duration of the bullet's flight. The bullet guides itself onto the point of aim by inbuilt sensors and fins which cause it to follow the laser beam to the target. The range and accuracy of this system is classified, but it does appear that it enables the shooter to hit anything he can see, every time.

If weapons can be developed to hit targets at ever-greater ranges, the optical aiming devices must improve commensurately. Modern scopes are produced to incredibly fine tolerances and have the ability to offer pretty much whatever magnification, or range of magnification, the user desires. The latest 'smart' scopes, however, use sensors and computers to allow for air pressure, temperature, humidity, azimuth and most of the effects bullets in flight have to contend with except for one – the wind. Wind is the most influential external factor affecting the flight of a bullet at any range. The shooter cannot stop the wind, obviously, so he surely has to estimate it the best he can and make allowances? Yet it seems that reasonably soon shooter or spotter estimation of wind will no longer be required. The Pentagon, through the Defense Advanced Research Projects Agency (DARPA), is currently running something called the 'One Shot' programme, which relates to achieving one-shot kills at extreme ranges despite the problems caused by variable winds. The agency states it is aware that 'a 10mph wind could produce a miss even at 400 meters while in other cases the deviation could be much worse, exceeding 3 meters at 1200 meters range'. To compensate for wind, the sniper spotter team have to make all manner of complex estimations, all approximate, while under great stress.

DARPA says that it wants a scope to work the wind factors out for itself and adjust the point of aim automatically. There are two contenders for the prize in this arena. The first idea, and simplifying a great deal, is to shoot thousands of laser beam flashes at the target in a split second. As the reflections of these flashes are monitored a profile of the eddies in the atmosphere between rifle and target is built up and the sights are adjusted accordingly. The alternative system takes a series of high-speed electronic photographs of the target and, because the eddies in the atmosphere distort the electromagnetic phase of the light waves in the photograph, the scope can use a series of algorithms to work out the wind effect and allow for them in the scope sight picture. It is expected that these two systems will boost the sniper's kill rate tenfold or more and allow snipers to engage targets in less than one second, though the reader might consider this system would be fairly irrelevant should guided bullets be issued.

In a further recent development, a Super Resolution Vision System (SRVS) claims it will not merely make snipers more accurate, but that it will make them functionally invisible as well. The idea is that the system uses heat haze, the shimmer which appears on a hot day over roads or in the distance, to help a sniper rather than hinder him, as it does at present. Heat haze is effectively a series of atmospheric 'lenses' that change over time as the air moves in the heat. SRVS takes a series of high-speed views of the area and using digital technology breaks down the effects and makes allowance for them. It makes one accurate picture from the accurate segments in a whole series of distorted pictures. The implications of this are quite astounding. Even if a target was looking right at a sniper through hot air, the heat haze would hide the sniper from the target but SRVS would allow the sniper to see the target perfectly and take that shot accurately through automatically corrected sights. Special forces snipers are expected to be field testing this kit in the very near future.

DARPA has awarded a $3.93 million contract to Lockheed Martin to develop a hi-tech rifle scope which will be of great help to designated marksmen initially, and eventually to all riflemen. The problem with optical sights at the moment is that they have to be set for range and wind and so on, which all seems to take a lot of time when someone is shooting at the user. The new scope, called the Dynamic Image Gunsight Optic (DINGO) combines a standard laser range-finder, a ballistic computer and some other sensors which, between them, focus the magnification of the scope optimally and move the crosshairs within the wide-angle viewing screen so that they lay on to the bullet's point of impact. Effectively, the shooter just has to point and shoot at any range up to about 600m (656yd) with excellent speed and accuracy. The first phase of this project will fit the system to M4 and M16 rifles and significantly increase both their accuracy and speed of use at variable ranges. DINGO may be a very useful invention, and increase the effectiveness of the infantry soldier by an order of magnitude, because shooting at speed in a contact situation has been known to produce poor marksmanship. Lockheed Martin recently applied the One Shot crosswind measurement technology mentioned above to a prototype spotter scope. They say field tests showed that snipers were able to engage targets twice as quickly, and double their probability of a first-round hit, at distances beyond 1,000m (1,094yd). Again, this would be rendered obsolete if a guided bullet were in use.

For all the undoubted improvements in development, some things never change, however. We have already seen that a rifle designed for long-range uses always fires a bullet at supersonic speed, and therefore that the bullet will make a loud supersonic announcement as it passes the target. There is nothing in development to reduce this, and the laws of physics mean that this noise signature is likely to remain for good. The propellant that fires the bullet also makes a bang and, though this can be reduced with a suppressor, it cannot be eliminated entirely, so electronics can find where the sniper is shooting from by listening. And finally all rifles make a flash when they fire. This is not noticeable to the human eye in daylight, but it is still there and can be seen by a suitable camera, which means that if someone has the equipment to detect it the sniper will find it difficult to hide after taking the shot. All three of these factors limit the use of a rifle as a sniping weapon severely. Or at least they limit the shots that can be taken while giving the shooter a reasonable chance of escaping alive. It would be a major advance for the sniper's survivability if he were able to take a shot and remain undetected by making no sound or flash.

Beyond the realms of propellant, cartridge and bullet, there are other developments in projectile propulsion that augur a radically different future. Admittedly, the medium-term applications of many of these technologies remain outside the utility of individual snipers, owing to size and weight considerations. Yet the impossibilities of today can become the actualities of the future, so it is worth touching on these technologies to understand where modern armies might be heading. One such development is known as the railgun.

A railgun fires projectiles, like any other gun, but rather than using a chemical explosive to blast them down a tube it uses magnetism. Technically it is called a homopolar or linear motor and works by having two rails connected only by the metallic projectile. A current is passed into one rail, through the projectile and back through the other rail. This causes an interaction called the Lorentz Force with the magnetic fields generate around the rails and pushes the projectile along like a magnetic train. Only on a railgun it pushes it much, much faster. The only downsides are that to power the equipment takes quite a lot of electricity and the projectile cannot contain explosive.

The French inventor Louis Octave Fauchon-Villeplee invented an electric cannon in 1918. He filed for a US patent, which was issued in July 1922 as patent no. 1,421,435 'Electric Apparatus for Propelling Projectiles'. The Germans designed an anti-aircraft weapon from this idea during World War II, but it was never built. After the war it was discovered that it would have worked and projected bullets at 2,000m/sec (6,561ft/sec), three times the speed of a conventional bullet, but it would have used enough electricity to power a fair-sized city. Experimentally, projectiles have been fired at speeds of up to 20,000m/sec (65,616ft/sec), but they tend to melt in the air like meteorites. And there are problems with arc welding the projectile to the rails and a few other considerations, but given a light enough power source in the future the railgun could be configured as a superlative sniping rifle. The latest research into lightweight railguns also suggests that they could be used as machine guns or rifles. As a machine gun a tremendous rate of fire could be achieved, as fast as projectiles could be fed into the machine. The expected projectile velocities being considered are around the 3,500m/sec (11,482ft/sec) mark which would make the projectile act like a cannon shell when it struck an object because of the incredible kinetic energy involved, very much like a meteor hitting the earth.

The Yugoslavian Military Technology Institute developed a railgun with 7kJ kinetic energy in 1985. In 1987 a successor was created, project EDO-1, which used a projectile with a mass of 0.7g (0.02oz) and achieved speeds of 3,000m/sec (9,842ft/sec). With a mass of 1.1g (0.04oz) it reached speeds of 2,400m/sec (7,874ft/sec) and used a track length of 0.7m (2ft 4in). According to those working on it, with other modifications it was able to achieve a speed of 4,500m/sec (14,763ft/sec). The aim was to achieve projectile speed of 7,000m/sec (22,965ft/sec). At the time, it was considered a military secret. This is not surprising because with such a short track length it was well on the way to being a deployable rifle-type weapon. Admittedly the size of the power supply would still be a significant issue.

In October 2006 the US Naval Surface Warfare Center Dahlgren Division demonstrated an 8MJ railgun firing 3.2kg (7lb) projectiles at a speed of 2,400m/sec (7,874ft/sec). This, the reader will notice, is more artillery weight projectile than rifle. Compare this velocity with the bullet from an M16 rifle, which travels at around 950m/sec (3,117ft/sec). This test bed is a prototype of a weapon that will be twice that power and deployed aboard navy warships. The main problem they have had with actually operating a railgun cannon system is that the gun's rails wear out quite quickly due to the immense heat produced by firing. The kinetic energy delivered by these weapons is sufficient to do more damage than a BGM-109 Tomahawk missile with a conventional warhead.

The US military is currently funding railgun experiments at the University of Texas at Austin Institute for Advanced Technology. Military style railguns capable of delivering tungsten armour-piercing bullets with kinetic energies of 9MJ have been developed. This is enough energy to deliver 2kg (4.4lb) of projectile at 3,000m/sec (9,842ft/sec), and at that velocity a rod of tungsten will act like a supercharged discarding sabot round and pass through several feet of armour, creating a huge detonation as it does so. On 31 January 2008, the US Navy tested a railgun that fired a projectile at 10.64MJ with a muzzle velocity of 2,520m/sec (8,267ft/sec). Its expected final performance is a muzzle velocity over 5,800m/sec (19,028ft/sec) and it will be accurate enough to hit a 5m (16ft) target over 370km (200 miles) away. And it can do that while firing ten shots per minute! The power comes from a bank of super-capacitors which store, and release in bursts, the electricity coming from a bank of generators. Even the most hardened sceptic must accept that the railgun concept is one of the front-runners for a long-range sniper weapon of the future, especially if the projectile could be guided with a laser as was discussed earlier. With the sort of muzzle velocities available, and laser guidance, one-shot kills against human targets at ranges of 16km (10 miles) and more are quite feasible.

An alternative development to the railgun is the high-energy laser. A laser is a beam of focused light in which all the light waves are polarized and march in step. This enables it to potentially travel great distances while carrying heat energy, without spreading out as a torch beam does. The idea of a laser as a weapon therefore is that it should carry enough heat energy to the target to do it serious damage. Unfortunately it is actually very difficult to make a laser with enough power to do anything useful aside from being a pointer. The first lasers were based around something like a neon light tube, in which electricity was pumped in and a laser beam came out. They were excellent for sending a focused beam of light to the moon, but they were not very powerful and could not be made powerful enough to do serious damage at a distance other than to someone's eyes. This is the type of laser which is still used as a toy, teaching pointer or aiming marker on a short-range firearm.

There followed the development of a whole host of chemical, gas and other types of laser, all capable of transmitting high levels of energy but at great production cost. Gas lasers were the best for weapons, because so much energy could be put into them, but their production required batteries of rocket engines burning poisonous gases and blasting the result through nozzles at speeds around six times the speed of sound. This made a laser light by doing something to the atoms in the gas and the light was caught by mirrors and directed away. A laser system of this type was actually installed in an aircraft the size of a Boeing 747, which it filled, and rigged up with the correct aiming systems it could shoot down a missile more than a mile away. But of course it was tricky to vent the gas exhaust and it soon ran out of fuel. Some of the other technical problems relating to poisoning and pollution were even worse to deal with.

The United States constructed this amazing machine and so, of course, the Soviets had to copy it because the Cold War was still raging. There was a rumour at the time that it formed the basis for the 'Star Wars' US defence system and was actually set up by President Reagan so that copying it would bankrupt the Russians. However, though a laser can be created that carries a weapons-grade amount of energy, the weapon's manufacturers could not let the generals get too carried away with the idea of improving and scaling up this chemical laser system, because it didn't work in rain, snow or fog or if there was dust in the air – all things that block or otherwise spoil lasers. So it was only much use at very high altitudes, like on the edge of space where the atmosphere was very clear, and for shooting down things like long-range ballistic missiles. On a battlefield it would be either only useable at short range or very unreliable at long range being dependent on the weather.

Eventually the chemical laser system was replaced by something called a solid-state laser, which is something like a larger version of the laser in a DVD reader. This system avoided the use of poisonous gas, and several other problems, and was reduced in size until it could be towed by first three and then one articulated truck, or mounted on a ship, of course. In this format it was developed to shoot down incoming mortars, artillery and missiles which it did very well subject to a lack of smoke, fog or heavy rain. At the time of writing, these laser systems are being considered as a replacement for the

US Phalanx and Dutch Goalkeeper Close-in Weapon System (CIWS), where a type of heavy Gatling Gun is mounted on a ship or truck and directed by radar to shoot down incoming artillery and missiles. The reason for the switch over to laser is that the solid-state laser only needs electricity, admittedly a great deal of electricity, to keep firing indefinitely whereas a 20mm Gatling gun firing 4,500 rounds a minute uses a whole truck load of ammunition every few minutes.

There are other versions of laser weapons developed by both the United States and Russia which are mounted on smaller aircraft than before, such as the Osprey, and which are set up to be used in the ground-attack role. But there is still a doubt that soldiers will ever see a laser rifle that doesn't need a thick flex leading back to a mains plug. The first candidate is a futuristic looking rifle called the PHaSR Laser Weapon, which has a lot more bark than bite. It has the ability to fire a beam which will blind, dazzle or lightly burn an enemy depending on the focus.

The next contender is called the TR3 and was developed by a partnership between the catchily named US firms Xtreme Alternative Defense Systems and SPA Defense. The TR3 can deliver a 3-watt green laser at targets up to 2.4km (1.5 miles) away when operated at night. This is enough to blind someone, but not do much more damage. It is powered by a re-chargeable lithium-ion battery which allows for up to 30 plus minutes of continuous firing. By now the reader will understand that lasers are not going to be personal weapons any time soon because of the huge amounts of electric power they require to do any real damage. Yet through the long process of technological evolution, it could be that destructive lasers eventually do find their way into the hands of individual soldiers.

Remotely Operated & Robotic Weapons

This final section concerns robots actually working as snipers. Robotic and remote-control systems are already fully integrated into modern weapons. It is now standard practice in the latest AFVs, such as the US Stryker Combat Vehicle, to keep the operator of the external machine gun tucked well down inside the armoured cabin rather than sitting with his upper body exposed behind the gun. From inside the vehicle he operates the weapon remotely using a joystick and video monitor, the technology allowing him to deliver precision fire with a greatly reduced risk to personal safety.

UAVs are perhaps the most visible modern example of remote-controlled warfare. A UAV is a light aircraft which has no pilot on board and is flown by relay teams of people sitting in a remote office and communicating with it via a satellite linkage. The control systems give similar control to that of a piloted aircraft and the weapons systems are effectively any bombs or missiles that might be carried by a manned aircraft. The main two main advantages of a UAV are that it can operate for extended periods because there is not a single pilot who needs to sleep, and having no human aboard means it carries no potential hostage to an enemy state. A UAV can actually stay in the air for weeks, if it refuels from a flying tanker, because there is no pilot fatigue at all owing to the flight team of about 65 people working in shifts. The only practical limit to the length of time it can stay over a target is that when it runs out of missiles it has to return to base for a rearm.

Remote-control warfare is also straying into the world of sniping. The first part of the sniper's task to become remotely operated will be the actual shooting. Or perhaps it would be more accurate to say the observing and shooting, as the two so often go together. There is no difficulty in producing a static stand of some kind upon which are mounted a remotely operated camera and rifle. If such a unit were built to be easily portable, perhaps in a stripped down form, then a sniper team or high-grade infantry unit could carry one into a sniping position covertly, build a hide for it and set it up to watch the enemy more or less indefinitely. The control would be by remote satellite linkage and the operators could be anywhere in the world. By working shifts they would never tire and they could stay in position until the batteries ran down. Perhaps a solar panel could recharge the batteries and give extreme endurance as on a sailing boat?

The Small Calibre Ultra Light (SCUL) weapon station built by Precision Remotes was developed as a follow-on to the TRAPS T-250D, which is currently in service with the US Marine Corps. This weapon mount weighs only 25kg (55lb) and can be mounted on a tripod, vehicle or an unmanned ground vehicle (UGV). The system supports several weapons including the M249 SAW, the M240B machine gun (with 400 rounds of ammunition) and the .50-cal M82A1M/M107 (Barret) anti-materiel sniper rifle. The normal range of optronics fitted includes a zoom CCD video camera and a thermal- or image-intensified sensor for night operations. The reader will note that this weapon is already in service with the military. A simple modification to the currently issued SCUL produces a sniper system with the endurance of a drone and which is immune to counter-sniper fire. The only risk to our troops is when they are placing it in position. Such a machine could mount guard over a position or a road indefinitely, gathering intelligence and when appropriate shooting its own sniper rifle or calling in fire from drones or artillery. Everything a sniper team can do, aside from moving position, with tireless efficiency and no risk to men.

The next stage in the development of the robot sniper will be mounting the machinery described above on a mobile chassis of some kind. This will allow it to make its own way across country and into a firing position. It will, however, not be able to do this in a clandestine manner like a human sniper. The technology already exists for small combat UGVs, with all the advantages of mission duration and operator safety outlined above. More than 40 countries around the world have examples on test. It seems likely that these vehicles are the future of military combat, partly because they are the only way to defeat the IED threat that forms the major problem in a counter-insurgency war and partly because they reduce the human casualties that cause the political problems in such a war. This IED resilience is an indisputable benefit because it is impossible to find or protect against all IEDs and their use leads eventually, through the loss of men, to a call for troops to be withdrawn from the area of operation and a war lost by any other name.

The Gladiator is a superb demonstration of what can be done with existing technology and which represents an introduction to tactical unmanned ground vehicles (TUGVs). It mounts an MMG and can keep up with an infantry section over all but the roughest ground, while offering a high firing position from which it provides supporting fire. It was issued to the Marines, but has recently been shelved, allegedly because they 'lost confidence' in it. The reality is that the Gladiator was withdrawn following political pressure, fuelled by public unease at the idea of a robot warrior. Such a machine set up for satellite control could mount a sniping rifle and move into position itself with, again, all the benefits of mission duration and operator safety provided by the UGV. It could also be helicopter deployed to a landing zone close to its intended firing position. On the downside, it could not build a hide, climb trees or move across country in a clandestine manner.

It is time now to take a look at very different form of mechanized sniper vehicle – the flying sniper. The Vigilante 502 rotor UAV is a modified version of the Ultrasport 496, which is nothing but an ultralight class kit helicopter. It is powered by a 115hp Rotax piston engine, the same as many ultralight aircraft, and can fly at 217km/h (135mph) for up to nine hours with a ceiling of 3,962m (13,000ft). UAVs like the Predator and Reaper are not only fixed-wing aircraft, with the concomitant requirement for a landing strip, but they are also designed to deliver area weapons and high explosives from great altitude, weapons that have the potential for causing a lot of collateral damage in a similar way to artillery or an air strike.

The Vigilante, however, can carry up to 172kg (380lb) of payload which is plenty for a light machine gun and ammunition or, as in this case, a sniper rifle. The rifle mounted is a .338in Lapua Magnum weapon similar to the latest British sniping rifle which has been making 2,500m (2,734yd) kills in recent times. It shoots out of a turret built by Space Dynamics Laboratory, and has both a situational awareness camera (wide-angle spotter) and a zoom telescopic sight. Control of this machine at present is by means of an Xbox 360 gaming controller with a 32km (20-mile) range, but it would not be difficult to equip it with satellite-transmitted control like the big fixed-wing UAVs.

To a certain extent the Vigilante fills the role mentioned above for a remotely operated sniping rifle that creeps into position. A small helicopter does make noise, but it can be silenced to some extent, and there is electronic equipment that can broadcast out-of-phase chopper sound to further silence its approach. Fitting this would enable a covert approach to a sniping position at night. This UAV can fire from the air with modest accuracy but, given how small choppers buck around, a shot from a landed position would be the way to eliminate a high-value target at long range.

There is a final class of robot warrior that many people find disturbing. TUGVs have been developed that consist of a high-speed armoured chassis mounting a mini gun and which effectively think for themselves. They are called autonomous tactical ground vehicles (ATGVs). An ATGV is typically given an area to patrol or search and it will follow roads or tracks or quarter the area like a search dog until it finds something it regards as an enemy. Then it will destroy that enemy. However, the application of an ATGV or *thinking* robot to the sniper role, and to the general combat role too, would only be an advantage where satellite control was not possible due to destruction of the control relay satellites or electronic jamming of some kind. This is because a human operator will be able to make finer and more reliable life-and-death decisions than a robot for some while to come.

So we see that the future of sniping could be transformed over the next decades. Today snipers make a personal observation of a target, often at great risk to themselves, and can make a precision kill out to ranges of several kilometres. The next generation of 'snipers' might be remote-controlled weapons platforms, deployed and controlled a long way from their operators, and capable of making kills at tens of kilometres with absolute precision. Either way, the fact remains that the sniper role is still integral to modern warfare.

CONCLUSION

If this book has proven one point, it is that sniping is one of the most technically, mentally and emotionally demanding of all the military skills. Technically, the sniper not only has to develop a deep-seated understanding of his weapon's mechanical and ballistic performance, but he also has to possess a firm practical grasp of the physics involved in putting a bullet onto a distant target. As we have seen, from the moment a bullet leaves the muzzle of a rifle, its trajectory is subject to a whole range of forces – gravity, wind, centrifugal spin, air temperature, humidity and so on. Trying to get that bullet to land precisely on a human figure at ranges of hundreds of metres requires a fusion of ingrained practical experience gained on the range or through combat, plus an understanding of the mathematical implications of each factor affecting bullet flight. Furthermore, the sniper must be capable of performing his calculations in a few seconds, the period in which a high-value target might cross between pieces of cover.

Alongside the technical challenges of sniping, the sniper needs the requisite mental toughness to cope with his frequently lonely vocation. The psychological make-up of special forces soldiers is rigorously assessed during the recruitment and training phases of their new military career, so those who go on to become snipers will already be tough-minded and resourceful individuals. Yet there is no doubt that being a sniper places additional burdens on even the most resilient warrior. A US sniper who served in the bush war in Angola during the 1970s here describes something of the sheer physical discomfort of hiding up and waiting for the shot:

> You had to hole-up in a pre-prepared hide, with somewhere to use for a latrine without exposing yourself and maybe sit it out for two days or more. Sometimes there were centipedes sharing the hole with me, as well as snakes, spiders and things with no names. I'd get bitten half to death; you could scratch or wiggle, or swear a long blue streak, especially in an exposed lay, and those long days baking in the tropical sun, feeding a wide variety of insect life, seemed eternal.

– Quoted in Pegler, p.219

Other sniping accounts tell similar tales of immobile discomfort under the elements, waiting with extraordinary patience for the target to expose himself or herself to the shot. After hours, sometimes days, of vigilance, the sniper finally pulls the trigger, and then the situation can change dramatically as the enemy returns volleys of counter-sniper fire and the sniper attempts to make his escape.

The final ingredient of the sniper experience is the emotional challenge that comes with the job. In the Introduction to this book, we noted how the sniper's profession has historically attracted a certain amount of alienation and even revulsion from his comrades. Whereas it has been perfectly acceptable for soldiers to kill in the fury and heat of battle, the sniper's considered killing of a human being has frequently isolated him from the general mass of soldiery. Thankfully, over recent decades the critical value of the sniper in warfare has become more apparent, and this has removed much of the stigma. Now long-range sniper kills become public news, and the men who take them (unless they are special forces soldiers, whose identities must remain concealed) are lauded for their skill.

Of course, any amount of acclaim doesn't remove the visceral nature of sniping. The sniper must observe a living, breathing human being, then put a high-velocity bullet into him, observing the individual's graphic death through the magnified viewpoint of a telescopic sight. A recent study by anthropologist Neta Bar also found that snipers tended not to dehumanize their victims as much as regular soldiers typically do. By observing their targets performing regular human actions, such as chatting with friends, eating, reading and washing, they were less able to view them simply as objects to be killed. One sniper in the study even watched a family weeping over the body of a man he had just killed. Yet, ironically, snipers also appear to suffer from a lower incidence of post-traumatic stress disorder (PTSD) than regular soldiers. Several explanations have been advanced for this apparently unlikely situation. One is that the sniper typically selects his targets consciously on the basis of their perceived threat to others, hence he is more able to justify the kill to himself. Second, snipers are self-selected for a lack of aversion to taking life in cold blood and this lack of aversion may be associated with a personality type less prone to PTSD. Finally, snipers are intelligent individuals, and combined with their sense of professionalism they quite often develop forms of self-counselling.

There are doubtless snipers who have been troubled in later life by the things that they have seen and done. The formation of modern programmes such as Sniper Post Operational Team Tactics for Emotional Recovery by the American Sniper Association attests that not all snipers are immune to PTSD. Nevertheless, the balance of evidence shows that snipers are individuals possessed of unique mental resources, who perform a job that is as difficult as it is invaluable. If recent conflicts have proved anything, it is that snipers remain utterly relevant to today's conflicts.

APPENDIX

The Armoury

The aim in this section is to present an overview of the relatively small number of genuine sniper rifles in use by the world's military today, rather than a list encompassing the hundreds of pseudo sniper weapons manufactured worldwide for amateur shooters. There is one representative example from each 'type' of sniping weapon. These include a classic bolt-action sniper rifle, a 12.7mm (.50in) anti-materiel rifle and a Russian semi-automatic weapon, suitable for the roles of both sniper and designated marksman. To complete the picture, this section ends with a description of some silenced and heavier weapons which are likely to be more widely adopted in the near future. True military sniping rifles are rarely less than 7.62mm in calibre, because a light bullet of smaller calibres loses momentum too quickly and is blown off course by the wind too easily to achieve a useful range. Therefore classic sniper rifles are between about 7.62mm and 9mm calibre, as this gives sufficient range without excessive weapon weight.

Most of the rifles intended for sniping are bolt-action to give maximum accuracy and because rate of fire for snipers is usually not a major issue. The British L115A3 Long Range Rifle described below is a good example of this class. Anti-materiel sniper rifles are designed to be used against light armour and have a calibre of 12.7mm (.50in), – for an example see the Barrett M107. In recent times there have been developed a number of still heavier-calibre sniping weapons, such as the 14.5mm and 20mm rifles. The 14.5mm rounds described below have roughly twice the power of the .50-cal round. These have been designed to defeat the ever-improving armour of armoured personnel carriers and similar vehicles, because many are now proof against the .50in Browning Machine Gun round. The heavier round which these weapons fire has also succeeded in extending the potential range of sniping weapons against human targets, and it is very likely that they will soon be used in that role.

Arctic Warfare Magnum: L115A3 Long Range Rifle

The Arctic Warfare Magnum is a superb rifle and has been nick-named 'Long Range Death' by soldiers of the British Army, owing to the many long-range hits it has achieved in combat. This is the weapon which made the longest ever confirmed sniper kill in 2009: 2,475m (2,707yd) by Corporal Craig Harrison. It is effectively the apogee of the classic bolt-action sniper rifle developed over the last 100 years and is probably close to the limit of what can be achieved with this style of weapon. The .338in L115A3 was designed to achieve a similar range to the .50in round with less weight and flash. Full metal jacket, hollow point and armour-piercing incendiary rounds are all available in this calibre.

Name: Arctic Warfare Magnum Brit: L115A3 Long Range Rifle
Manufacturer: Accuracy International (UK)
Designed range: 1,500m (1,640yd)
Calibre: .338in Lapua Magnum
Magazine capacity: 5
Action: Bolt-action
Sights: Schmidt & Bender 5–25x 56 PM II LP/Military Mk II 5–25x 56 0.1 MIL RAD parallax, illumination, double turn telescopic sights
Suppressor: Yes
Muzzle brake: Yes

Weight (without scope): 6.0kg (13.2lb)
Barrel length: 686mm (27in) stainless steel fluted
Overall length: 1,230mm (48.4in)
Stock: Folding
Trigger weight: Two-stage trigger; adjustable pull weight of 1–2kg (2.2–4.4lb)
Muzzle velocity: 936m/sec (3,070ft/sec)
Chamber pressure: 4,699kg/cm² (66,833psi)
Rifling: 1 in 279.4mm (50.4in) right-hand twist
Users: Bangladesh, Germany, Indonesia, Italy, Netherlands, Poland, Russia,
    United Kingdom

Barrett M107

The Barrett M107 came about through the US military's idea of using the .50in Browning round as a sniper cartridge. The Barrett came into service during the early 1980s and has since appeared in a number of incarnations with different names. It remains substantially the same rifle and is extremely effective against human beings and most light armour out to a range of nearly 2km (3.2 miles). An advantage of its semi-automatic mechanism is that several human targets can be engaged before they have time to move. One important point to remember is that the operator must not use the .50in sabot round or the rifle will explode. Though no body armour can stop this round, the armour of some modern light armoured vehicles can. The Barrett is slightly more powerful than the Russian 12.7mm weapons, but not as powerful as their 14.7mm rifles.

Name: M107
Manufacturer: Barrett (USA)
Designed Range: 1,800m (1,969yd) against human targets, more against vehicles
Calibre: 12.7mm (.50in) Browning Machine Gun; other loads are available
Magazine capacity: 10
Action: Recoil-operated; semi-automatic
Sights: Leupold 4.5x 14 Vary X
Suppressor: Yes
Muzzle brake: Yes
Weight (with scope): 12.9kg (28.4lb) empty
Barrel length: 737mm (29in)

Overall length: 1,448mm (57in)
Stock: Rigid
Trigger weight: N/A
Muzzle velocity: 853m/sec (2,799ft/sec)
Chamber pressure: N/A
Rifling: 1 in 381mm (15in)
Users: Australia, Bahrain, Belgium, Bhutan, Botswana, Brazil, Chile,
Czech Republic, Denmark, Finland, France, Georgia, Germany,
Greece, India, Israel, Italy, Jordan, Kuwait, Lebanon, Lithuania, Malaysia,
Mexico, Netherlands, Norway, Oman, Philippines, Poland, Portugal, Qatar,
Saudi Arabia, Singapore, Spain, Sweden, Turkey, United Arab Emirates,
United Kingdom, United States

Accuracy International AS50

The AS50 is one of a range of weapons designed to fire the Browning .50-cal machine-gun round or the more accurate specialist sniper loads of that design and calibre. It is relatively light, reliable and strips for carrying very quickly, unlike many other .50-cal sniping rifles. The design specifications also mean it will be able to fire the improved .50in cartridge which is currently being developed, to achieve even greater accuracy.

Name: AS50 sniper rifle
Manufacturer: Accuracy International (UK)
Designed range: 1,500–2,000m (1,640–2,187yd) normal
Calibre: 12.7mm (.50in)
Magazine Capacity: 5 or 10
Action: Gas-operated; semi-automatic
Sights: Mil spec Mk II in 6x, 10x, 3-12V and 4-16V
Suppressor: Yes
Muzzle brake: Yes
Weight (with scope): 15kg (33lb) with bipod
Barrel length: 692mm (27in)
Overall length: 1,369mm (54.8in)
Stock: Folding
Trigger weight: 1.5–1.8kg (3.3–4lb)
Muzzle velocity: 850m/sec (2,788ft/sec)

Chamber pressure: N/A
Rifling: N/A
Users: United Kingdom

XM2010 Enhanced Sniper Rifle

The XM2010 Enhanced Sniper Rifle is the new US general-issue sniper weapon, as opposed to the designated marksman weapon, and was designed to replace the earlier sniper rifle system, the M24. (The XM2010 was formerly known as the M24 Reconfigured Sniper Weapon System and is effectively an up-rated M24.) SEALs and other experienced operators have complained that this rifle is neither a good long-range sniping weapon, as it does not shoot far enough, nor a good designated marksman weapon due to its slow rate of fire. They have also noted that their shoulders were black and blue after firing the weapon, as a result of its harsh recoil. At the time of writing, modifications to the round are being tested to increase the effective range.

Name: XM2010 Enhanced Sniper Rifle
Manufacturer: Remington Arms
Designed range: 1,200m (1,312yd)
Calibre: .300in Winchester Magnum
Magazine capacity: 5
Action: Bolt
Sights: Leupold Mark 4 6.5–20x 50mm ER/T M5 or AN/PVS-29 clip-on sniper night sight
Suppressor: Yes
Muzzle brake: Yes
Weight (with scope): 7.95kg (17.5lb) loaded
Barrel length: 610mm (24in)
Overall length: 1,135mm (44.6in)
Stock: Adjustable length and cheek piece
Trigger weight: 1.6–2.3kg (3.5–5lb)
Muzzle velocity: 869m/sec (2,851ft/sec) using Mk 248 MOD 1 sniper load
Chamber Pressure: 4,781kg/cm^2 (68,000psi)
Rifling: 1 in 254mm (10in) right-hand twist
Users: United States

Dragunov SVD

A Russian gentleman by the name of Yevgeni Dragunov first thought up the principle of the 'designated marksman', which has now been adopted by the US military. A designated marksman within an infantry unit is required to increase the lethal range of the patrol armed with short-range assault rifles by hitting targets out to about 600–1,000m (656–1,094yd). To put his theory into practice, Dragunov developed the semi-automatic Dragunov rifle, which is accurate out to about 1,300m (1,422yd) when supplied with high-quality ammunition. For convenience it also fires a 7.62 x 54mm round already carried by the squad PKM machine-gun team. Western designated marksman weapons in the United States tend to be more accurate versions of the standard squad weapon with better sights. These weapons do not give the range of the Dragunov, but they allow a very high rate of fire and the sharing of rifle ammunition amongst all the squad.

Name: Dragunov SVD
Manufacturer: Izhmash Ordnance Factories, Izhevsk
Designed range: 1,300m (1,422yd)
Calibre: 7.62mm
Magazine capacity: 10
Action: Gas-operated; semi-automatic
Sights: PSO-1 telescopic sight
Suppressor: Primitive flash suppressor
Muzzle brake: No
Weight (with scope): 4.3kg (9.5lb) empty
Barrel length: 610mm (24in) shorter on some models
Overall length: 1,225mm (48in)
Stock: Folding on some models
Trigger weight: N/A
Muzzle velocity: 830m/sec (2,727ft/sec)
Chamber pressure: N/A
Rifling: N/A
Users: Afghanistan, Albania, Bangladesh, Belarus, Bulgaria, China (Norinco-made copy), Czech Republic, Finland, Georgia, Hungary, India (built under licence), Iran (locally produced as the Nakhjir Sniper Rifle), Iraq (variant based on the

SVD, known as the Al Kadesiah), Kazakhstan, Kyrgyzstan, Libya, Nicaragua, North Korea, Poland (the Polish SVD is known as the SWD-M and updated with a heavier barrel), Romania (built under licence), Russian Federation, Slovakia, Syria, Tajikistan, Turkey, Turkmenistan, Ukraine, Uzbekistan, Venezuela, Vietnam

Sniper Rifle with PSO-1 Scope

This superb, almost silent, weapon was developed by the Russians for clandestine work and urban warfare. It is issued to the Russian Spetsnaz and Ministerstvo Vnutrennikh Del (MVD; Ministry of Internal Affairs) troops. They used it with great success in the 2008 South Ossetia war, the Soviet occupation of Afghanistan and the various Chechen Wars. The Russians have another weapon with similar characteristics called the VSK-94.

Name: VSS Vintorez
Manufacturer: Tula Arms Plant Russia
Designed Range: 400m (437yd)
Calibre: 9 x 39mm
Magazine capacity: 10 or 20
Action: Gas-operated; semi-automatic
Sights: Standard PSO-1 scope
Suppressor: Silencer
Muzzle brake: No
Weight (with scope): 4.7kg (10lb)
Barrel length: 200mm (7.9in)
Overall length: 894mm (35in)
Stock: Wood or plastic
Trigger weight: N/A
Muzzle velocity: 290m/sec (951ft/sec)
Chamber pressure: N/A
Rifling: N/A
Users: Russia and client states

Heckler & Koch HK MP5SD

Since being used by the SAS in the 1980 Iranian Embassy siege in London, variants of the HK MP5 submachine gun have been adopted by special forces, police and other units worldwide. Though first developed in 1966, the MP5 range has been improved and modified in many ways. It remains the most popular submachine gun in the world, and though it has recently been superseded by the HK UMP it remains in production and widespread use purely due to its popularity. The silenced version of the MP5, the MP5SD, has a special barrel/silencer combination that bleeds off gas early from the barrel to slow the bullets from their normal supersonic speed to a sub-sonic velocity before they exit the barrel. This allows the use of normal issue 9mm ammunition rather than a special low-power load.

Name: HK MP5
Manufacturer: Heckler & Koch (Germany)
Designed Range: 200m (219yd)
Calibre: 9mm
Magazine capacity: 15–30
Action: Roller-delayed blowback; semi-/full-auto
Sights: Various
Suppressor: MP5SD (SD = *Schalldämpfer* – German for 'sound suppressor')
Muzzle brake: No
Weight (with scope): 3.5kg (7.7lb)
Barrel length: 146mm (5.7in) (MP5SD)
Overall length: 805mm (32in)
Stock: Folding
Trigger weight: N/A
Muzzle velocity: 315m/sec (1,033ft/sec)
Chamber pressure: N/A
Rifling: N/A
Users: Special forces and tactical police units worldwide

Istiglal IST

The Istiglal was developed in 2008 by Azerbaijan, but only revealed in 2009. It is a superb, reliable weapon with sufficient accuracy to hit a human target at 3,000m (3,281yd). More significantly, it will destroy any light armour or military helicopter now in production.

The author has predicted the emergence of a 14.5mm sniper weapon for some years, owing to the availability of the excellent Soviet bloc 14.5mm round, and the call for heavier anti-materiel weapons. The development of this sniper weapon from the machine gun that fires this ammunition is a direct parallel of the development of the .50-cal sniper rifles from the Browning M2HB. It is very likely that there will be high-accuracy ammunition and special effects rounds available very shortly.

Name: Istiglal IST
Manufacturer: Azerbaijani Defence Industry (Azerbaijan)
Designed range: 3,000–4,000m (3,281–4,374yd)
Calibre: 14.5mm
Magazine capacity: 10
Action: Recoil-operated; semi-auto
Sights: Pospí 8x 42D
Suppressor: Yes
Muzzle brake: Yes
Weight (with scope): 19.8kg (43.6lb)
Barrel length: 1,220mm (48in)
Overall length: 2,015mm (79in)
Stock: N/A
Trigger weight: N/A
Muzzle velocity: 1,132m/sec (3,713ft/sec)
Chamber pressure: N/A
Rifling: N/A
Users: Azerbaijan, Jordan, Pakistan, Turkey; Belarus, Georgia, Israel, Russia and Ukraine have already shown interest

Mechem NTW-20 14.5mm

The Mechem was designed for the open plains of South Africa and has a tremendous range and hitting power from the 14.5mm round. There is also a 20mm (0.8in) barrel, and change between the two barrels can be performed in about one minute. The purpose of this heavier-calibre barrel is that though the 20mm projectiles do not have the range of the 14.5mm, they are more efficient for carrying explosive and incendiaries.

Name: Mechem NTW-20
Manufacturer: DENEL Group (South Africa)
Designed Range: 2,300m (2,515yd) +
Calibre: 14.5mm
Magazine capacity: 3
Action: Bolt-action
Sights: 8x mag, long eye-relief, telescopic sight on quick detachable mount
Suppressor: Yes
Muzzle brake: Yes
Weight (with scope): 29kg (64lb)
Barrel length: 1,220mm (48in)
Overall length: 2,015mm (79in)
Stock: N/A
Trigger weight: N/A
Muzzle velocity: N/A
Chamber pressure: N/A
Rifling: N/A
Users: Several countries showing interest

20mm 'Take-Down Rifle'

It is strange that the US military has not officially adopted any weapons of larger calibre than 12.7mm (.50in) as sniper or anti- materiel weapons, given there is a definite requirement for use against improving armour and there are several overseas models already in production. Perhaps this is because the United States does not have a 14.5mm-calibre machine gun. This rifle is one of the several very heavy-calibre rifles manufactured in the United States, but not adopted by the military.

Name: 20mm 'Take-Down Rifle'
Manufacturer: Anzio Ironworks (United States)
Designed range: 3,000m (3,281yd)
Calibre: 20mm
Magazine capacity: 3
Action: Bolt-action
Sights: 32 /64x magnification
Suppressor: Yes

Muzzle brake: Yes
Weight (without scope): 17.7kg (39lb)
Barrel length: 245mm (9.6in)
Overall length: 2,032mm (80in)
Stock: N/A
Trigger weight: N/A
Muzzle velocity: 1,000m/sec (3,280ft/sec)
Chamber pressure: N/A
Rifling: N/A
Users: No military users

GLOSSARY

ACP: Automatic Colt Pistol (cartridge).

Ball: A standard, inert military bullet type, typically of brass jacketed lead.

Blowback: A system of firearms operation that uses the breech pressure generated upon firing to operate the bolt.

Bolt: The part of a firearm that closes the breech of the firearm and often performs the functions of loading, extraction and (via a firing pin) ignition.

Breech: The rear end of a gun barrel.

Breech-block: A mechanism designed to close the breech for firing; roughly analogous to 'bolt' but usually referring to an artillery weapon.

Bullet drop: The fall of a bullet, under the force of gravity, beginning as soon as it leaves the barrel of the gun.

Carbine: A shorter model rifle.

Cartridge: A single unit of ammunition containing the bullet, propellant, primer and case.

Chamber: The section at the rear of the barrel, usually of greater diameter than the barrel, into which the cartridge is seated for firing.

Cheek rest: A raised section of the stock on which the shooter rests his cheek to obtain the correct eye-relief and sight picture through a telescopic sight.

Closed bolt: Refers to firearms in which the bolt/breech-block is closed and locked up against the chamber before the trigger is pulled.

Cook-off: The involuntary discharge of a cartridge by the build-up of heat in the chamber from previous firing.

Crosshairs: Intersecting vertical and horizontal lines within the optics of a telescopic sight, used to aim a sniper rifle and to judge range and deflection.

Deflection: The distance of aim in front of a moving target so that the bullet strikes the target at a predetermined point after its flight.

Delayed blowback: A blowback mechanism in which the recoil of the bolt is mechanically delayed while the chamber pressures drop to safe levels.

Ejector: The mechanism that throws an empty cartridge case clear of a gun following extraction from the chamber.

Extractor: The mechanism that removes an empty cartridge case from the chamber after firing.

Eye-relief: The distance between the eye of the shooter and the rear lens of the telescopic sight.

Field of view: The visual extent of an image seen through an optical device, such as binoculars or a telescopic sight.

Gas operation: A system of operating the cycle of a firearm using gas tapped off from burning propellant.

Graticule: A variant name for crosshairs.

Line of sight: A direct sight relationship between the shooter and the target.

Lock time: The time interval between pulling the trigger and the gun firing.

Mean point of impact: Essentially the geometric centre of a group of shots fired at a target; the MPI is used to zero the sight for accurate shooting.

Minute of angle: 1 minute of angle (MOA) almost exactly equals 1in at 100yd, 2in at 200yd and so on. (In metric 1 MOA at 100m equals 2.9cm.) Telescopic sight adjustments work in MOA.

Mount: The physical means of attaching a telescopic sight to the body of a rifle.

Muzzle brake: A device fitted to the muzzle of a large-calibre firearm, which deflects propellant gases to the side and rear and therefore helps reduce felt recoil.

Muzzle velocity: The speed of the bullet as it leaves the muzzle of the gun. Note that the velocity of the bullet drops significantly once it has left the bore.

Open bolt: Refers to firearms in which the bolt/breech-block is held back from the breech before the trigger is pulled.

Range-finder: An optical device that performs precise mechanical or computerized range calculations.

Receiver: The main outer body of a gun, which holds the firearm's action or 'working parts'.

Recoil operated: An automatic weapon powered through the extraction, ejection and loading cycles by the forces of recoil. In a short-recoil weapon, the barrel and bolt recoil for less than the length of a cartridge before they unlock and ejection takes place.

Reticle: Another name for crosshairs.

Semi-automatic: A weapon that fires one round and reloads ready for firing with every pull of the trigger.

Sub-sonic: Ammunition that travels below the speed of sound, thereby avoiding the supersonic crack that forms a major sound component of a regular military rifle round in flight.

Suppressor: Device to reduce the sound of the burnt propellant gas leaving the muzzle during firing.

Tracer: A bullet incorporating a chemical compound that ignites when the cartridge is fired, enabling the shooter or other observer to observe the flight of the bullet.

Trajectory: The path taken by a bullet during its flight from the muzzle to the target.

Windage: Normally this refers to the effect of lateral forces on the flight of a bullet, typically the wind, but historically also to the gap between bullet and bore in a musket.

Zeroing: The process of ensuring that the point of aim indicated by the sights is the point at which the bullet will strike a horizontal target in calm air.

BIBLIOGRAPHY

Chandler, N.A and R.F. Chandler, *Death from Afar*, (Iron Brigade
Armory Publishing, NC, 1993)

Chandler, Roy F.,*White Feather: Carlos Hathcock USMC Scout
Sniper: An Authorized Biographical Memoir* (Iron Brigade
Armory Publishing, NC, 1997)

Dockery, Kevin, *Stalkers and Shooters: A History of Snipers* (London,
Penguin, 2007)

Dougan, Andy, *The Hunting of Man: A History of the Sniper* (London,
Fourth Estate, 2004)

Gilbert, Adrian, *Sniper: The Skills, the Weapons, and the Experiences*
(New York, St Martin's Paperbacks, 1996)

Henderson, Charles, *Marine Sniper: 93 Confirmed Kills* (New York,
Berkley Books, 2001)

Henderson, Charles, *Silent Warrior* (New York, Berkley Books, 2003)

Hesketh-Pritchard, H.V., *Sniping in France 1914–18: With Notes on
the Scientific Training of Scouts, Observers, and Snipers* (London,
Leo Cooper, 1994)

Law, Clive M., *Without Warning: Canadian Sniper Equipment of
the 20th Century* (Ottawa, ON, Service Publications, 2005)

Pegler, Martin, *Out of Nowhere: A History of the Military Sniper*
(Oxford, Osprey Publishing, 2011)

Plaster, Maj. John L., *The History of Sniping and Sharpshooting*
(Boulder, CO, Paladin Press, 2007)

Roberts, Craig and Charles W. Sasser, *Crosshairs on the Kill Zone: American Combat Snipers, Vietnam Through Operation Iraqi Freedom* (New York, Simon and Schuster, 2004)

Sakaida, Henry, *Heroines of the Soviet Union 1941–45* (Oxford, Osprey Publishing, 2003)

Sasser, Charles and Craig Roberts, *One Shot, One Kill* (New York, Pocket Books, 1990)

Senich, Peter R., *The One-Round War: USMC Scout-Snipers in Vietnam* (Boulder, CO, Paladin Press, 1996)

Senich, Peter R., *The German Sniper, 1914–1945* (Boulder, CO, Paladin Press, 1982)

Senich, Peter R., *The Complete Book of U.S. Sniping* (Boulder, CO, Paladin Press, 1988)

Shore, C., *With British Snipers to the Reich* (London, Greenhill, 1997)

INDEX

LIKED SPECIAL FORCES SNIPER SKILLS?

Then you will enjoy similar titles from Osprey Publishing...

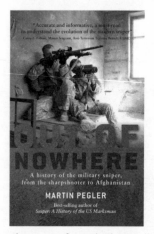

OUT OF NOWHERE: A HISTORY OF THE MILITARY SNIPER, FROM THE SHARPSHOOTER TO AFGHANISTAN

The sniper is probably the most feared specialist warrior and the most efficient killer on the battlefield. Endlessly patient and highly skilled, once he has you in his crosshairs, your chances of survival are slim. This revised edition of *Out of Nowhere* provides a comprehensive history of the sniper, giving insights into all aspects of his life; his training tactics, equipment and the psychology of sniping are examined in the context of the major wars of modern times – including the American Civil War, both world wars, the Vietnam War and the ongoing conflict in Afghanistan. First-hand accounts from veteran snipers demonstrate their skill and extraordinary courage and show why they are still such a vital part of any war.

£8.99
US $ 14.95
CAN $16.95

ALSO AVAILABLE AS AN EBOOK

www.ospreypublishing.com